ASTRONOMY AND
ASTROPHYSICS LIBRARY

Norman K. Glendenning

# Special and General Relativity

With Applications to White Dwarfs,
Neutron Stars and Black Holes

First Edition

 Springer

Norman K. Glendenning
Laurence Berkeley National Laboratory
Nuclear Science Division & Institute for
Nuclear & Particle Astrophysics
One Cyclotron Rd, MS 70R319
Berkeley, CA 94720
U.S.A.

ISBN  978-1-4419-2366-0          eISBN 10: 0-387-47109-X
                                 eISBN 13: 978-0-387-47109-9

Printed on acid-free paper.

Printed in the United States of America.

9 8 7 6 5 4 3 2 1

springer.com

I dedicate this book to my youngest child, Nathan

# Preface

There are magnificent tomes on General Relativity, the greatest of all being *Gravitation* by C. Misner, K. Thorne, and J. Wheeler, and *Gravitation and Cosmology* by S. Weinberg, both of which include some of the interesting historical context. Nevertheless, a smaller book that the covers the mathematics of curved spacetimes through to the development of the General Relativity—briefly but completely—and provides as well succinct chapters on White Dwarfs, Neutron Stars, Black Holes, and Cosmology, has been suggested by some who have read my earlier book on *Compact Stars*. I would like to mention as well a little known book by the great Paul Dirac. It will probably hold the distinction of being the briefest of all expositions of General Relativity alone. There are no applications to relativistic objects aside from black holes.

*Norman K. Glendenning*
Lawrence Radiation Laboratory
One Cyclotron Road
University of California
Berkeley, California 94720

January 1, 2007

# Contents

# 1

# Introduction

"In the deathless boredom of the sidereal calm we cry with
regret for a lost sun ..."
*Jean de la Ville de Mirmont, L'Horizon Chimérique.*

Compact stars—neutron stars and white dwarfs—are the ashes of luminous stars. One or the other is the fate that awaits the cores of most stars after a lifetime of millions to billions of years. Whichever of these objects is formed at the end of the life of a particular luminous star, the compact object will live in many respects unchanged from the state in which it was formed. Neutron stars themselves can take several forms—hyperon, hybrid, or strange quark star. Likewise white dwarfs take different forms though only in the dominant nuclear species. A black hole is probably the fate of the most massive stars, an inaccessible region of spacetime into which the entire star may fall at the end of the luminous phase.

Neutron stars are the smallest, densest stars known. Like all stars, neutron stars rotate—some as many as a few hundred times a second. A star rotating at such a rate will experience an enormous centrifugal force that must be balanced by gravity else it will be ripped apart. The balance of the two forces[1] informs us of the lower limit on the stellar density. Neutron stars are about $10^{14}$ times more dense than the average Earth density. Some neutron stars are in binary orbit with a companion. Application of orbital mechanics allows an assessment of masses in some cases. The mass of a neutron star is typically 1.4 - 1.5 solar masses, though some may be heavier: their radii are about ten kilometers. Into such a small object, the entire mass of our sun and more, is compressed.

We infer the existence of neutron stars from the occurrence of supernova explosions (the release of the gravitational binding of the neutron star) and observe them in the periodic emission of pulsars. Just as neutron stars acquire high angular velocities through conservation of angular momentum, they acquire  strong magnetic fields through conservation of magnetic flux

---

[1] $GMm/R^2 \geq m\Omega^2 R \longrightarrow G\bar{\rho} \geq \frac{3}{4\pi}\Omega^2$ ($\bar{\rho}$ = av. Density, $\Omega$ = angular frequency

during the collapse of normal stars. The two attributes, rotation and strong magnetic dipole field, are the principle means by which neutron stars can be detected—the beamed periodic signal of pulsars.

The extreme characteristics of neutron stars set them apart in the physical principles that are required for their understanding. All other stars can be described in Newtonian gravity with atomic and low-energy nuclear physics under conditions essentially known in the laboratory.[2] Neutron stars in their several forms push matter to such extremes of density that nuclear and particle physics—pushed to their extremes—are essential for their description. Further, the intense concentration of matter packed into neutron stars with a typical radius of 10 km makes them fully relativistic stars describable only in General Relativity, Einstein's theory of gravity which describes the way the weakest force in nature arranges the distribution of the mass and constituents of the densest objects in the universe.

## 1.1   Compact Stars

"With all reserve we advance the view that supernovae represent the transition from ordinary stars into *neutron stars*, which in their final stages consist of closely packed neutrons."
*W. Baade and F. Zwicky, 1933.*

Stars like our Sun are, in the words of Sir Arthur Eddington, great furnaces that will burn for 12 billion years. In the late stages of burning, the Sun, as with all low-mass stars, will expel some of its outer layers of gas to form a nebula. Meanwhile, the core will sink into a *white dwarf* stage, a hot and dense object with no internal source of energy. It will cool slowly for billions of years and in time may even become crystalline.

More massive stars, up to about 5 solar masses ($5M_\odot$), will end their lives in a supernova explosion, which casts off most of the star. Under the attraction of gravity the inner shells will fall in a fraction of a second to form a dense *neutron star* about 20 kilometers across. The birth of a neutron star, if close enough—as in the case of the explosion of 1987 in the Large Magellanic Cloud—is signaled by an intense neutrino shower in addition to the light display. The explosion may illuminate the sky for many weeks before finally fading, as in the case of the 1054 event—the Crab Supernova. More massive stars, at the end of their luminous lives, sink into the oblivion of a black hole.

---

[2]Luminous stars evolve through thermonuclear reactions. These are nuclear reactions induced by high temperatures but involving collision energies that are small on the nuclear scale. In some cases the reaction cross-sections can be measured with atomic accelerators, while in others, measured cross-sections must be extrapolated to lower energy.

The notion of a neutron star made from the ashes of a luminous star at the end point of its evolution goes back to 1933 and the study of supernova explosions by the German-born Walter Baade and the Swiss-born Fritz Zwicky working in astrophysics at the California Institute of Technology [1].

The neutron had only been discovered a mere two years prior to Baade and Zwicky's insightful interpretation of supernovae. James Chadwick, working at the Cavendish Laboratory at Cambridge University discovered this electrically neutral cousin of the proton. Astrophysics was destined to change forever when neutron stars were eventually discovered by Joselyn Bell and Anthony Hewish at Cambridge in 1967 [2]. J. Robert Oppenheimer and G. M. Volkoff had already devised a theoretical model for such a star almost 30 years prior to the actual discovery [3].

We use "neutron star" in a generic sense to refer to stars as compact as described above. How does a star become so compact as neutron stars and why is there little doubt that they are made of many types of baryons and possibly have a core of quarks—the constituents of baryons? These are questions that we explore.

During the luminous life of a star, part of the original hydrogen is converted in fusion reactions to heavier elements by the heat produced by gravitational compression. When sufficient iron—the end point of exothermic fusion—is made, the core containing this heaviest ingredient collapses and an enormous energy is released in the explosion of the star. Baade and Zwicky guessed that the source of such a magnitude as makes these stellar explosions visible in daylight and for weeks thereafter must be gravitational binding energy. This energy is released by the solar mass core as the star collapses to densities high enough to tear all nuclei apart into their constituents. The most famous such explosion was the one that produced the beautiful Crab Nebula. It was reported by the court astronomer to the emperor of the Sung dynasty in 1054:—"I bow low. I have observed the apparition of a guest star ... its color was an iridescent yellow ...". The nebula is now about 10 light years across, and growing still, powered by the a rapidly rotating Crab pulsar The power emitted by the rotating magnetized pulsar is equal to 100,000 times the power emitted by our Sun.

By a simple calculation one learns that the gravitational energy acquired by the collapsing core is more than enough to power such explosions as Baade and Zwicky were detecting. Their view as concerns the compactness of the residual star has since been supported by many detailed calculations, and most spectacularly by the supernova explosion of 1987 in the Large Magellanic Cloud, a nearby minor galaxy visible in the southern hemisphere. The pulse of neutrinos and photons observed in several large detectors carried the evidence for an integrated energy release over $4\pi$ steradians of the expected magnitude.

The gravitational binding energy of a neutron star is about 10 percent of its mass. Compare this with the nuclear binding energy of 9 MeV per nucleon in iron which is one percent of its mass. We conclude that the release of gravitational binding energy at the death of a massive star is of the order ten times greater than the energy released by nuclear fusion reactions during the entire luminous life of the star. The evidence that the source of energy for a supernova is the binding energy of a compact star— a neutron star—is compelling. How else could a tenth of a solar mass of energy be generated and released into the heavens in such a short time?

Neutron stars are denser than was thought possible by physicists at the turn of the century. At that time astronomers were grappling with the thought of white dwarfs whose densities were inferred to be about a million times denser than the earth. It was only following the discovery of the quantum theory and Fermi-Dirac statistics that very dense *cold* matter—denser than could be imagined on the basis of atomic sizes—was conceivable.

Prior to the discovery of Fermi-Dirac statistics, the high density inferred for the white dwarf Sirius seemed to present a dilemma. For while the high density was understood as arising from the ionization of the atoms in the hot star making possible their compaction by gravity, what would become of this dense object when ultimately it had consumed its nuclear fuel? Cold matter was known only in the atomic form it is on earth with densities of a few grams per cubic centimeter. The great scientist Sir Arthur Eddington surmised for a time that the star had "got itself into an awkward fix"—that it must some how re-expand to matter of familiar densities as it cooled, but it had no remaining source of energy to do so.

The perplexing problem of how a hot dense body without a source of energy could cool persisted until R. H. Fowler "came to the rescue"[3] by showing that Fermi-Dirac degeneracy allowed the star to cool by remaining comfortably in a previously unknown state of cold matter, in this case a degenerate[4] electron state. A little later Baade and Zwicky conceived of a similar degenerate state as the final resting place for nucleons after the supernova explosion of a luminous star.

The constituents of neutron stars — leptons, baryons and quarks — are degenerate. They lie helplessly in the lowest energy states available to them. They must. Fusion reactions in the original star have reached the end point for energy release—the core has collapsed, and the immense gravitational energy—converted to neutrinos and photons—has been carried away. The star has no remaining source of energy to excite the Fermions. Only the

---

[3]Eddington in an address in 1936 at Harvard University.

[4]Nucleons and electrons obey the Pauli exclusion principle, according to which each particle must occupy a different quantum state from the others. A degenerate state refers to the complete occupation of the lowest available energy states. In that event, no reaction and therefore no energy generation is possible—hence the name —degenerate state.

Fermi pressure and the short-range repulsion of the nuclear force sustain the neutron star against further gravitational collapse—sometimes. At other times the mass is so concentrated that it falls into a black hole, a dynamical object whose existence and external properties can be understood only in the Theory of General Relativity.

## 1.2    Compact Stars and Relativistic Physics

Classical General Relativity is completely adequate for the description of neutron stars, white dwarfs, and the exterior region of black holes as well as some aspects of the interior.[5] The second chapter briefly reviews Special Relativity and the next is devoted in detail to the mathematics of curved spacetimes and then to General Relativity. The goal is to rigorously arrive at the equations that describe the structure of relativistic stars—the Oppenheimer–Volkoff equations—the form taken by Einstein's equations for spherical static stars. Two important facts emerge immediately. No form of matter whatsoever can support a relativistic star above a certain mass called the limiting mass.[6] Its value depends on the nature of matter but the existence of the limit does not. The implied fate of stars more massive than the limit is that either mass is lost in great quantity during the evolution of the star or it collapses to form a black hole.

Black holes—the most mysterious objects of the universe—are treated at the classical level and only briefly. The peculiar difference between time as measured at a distant point and on an object falling into the hole is discussed. And it is shown that within black holes there is no statics. Everything at all times must approach the central singularity. Unlike neutron stars and white dwarfs, the question of their internal constitution does not arise at the classical level. They are enclosed within a horizon from which no information can be received. The ultimate fate of black holes is unknown.

Luminous stars are known to rotate because of the Doppler broadening of spectral lines. Therefore their collapsed cores, spun up by conservation of angular momentum, may rotate very rapidly. Stellar rotation—its effects on the structure of the star and spacetime in the vicinity, the limits on rotation imposed by mass loss at the equator and by gravitational radiation, and the nature of compact stars that would be implied by very rapid rotation—is discussed in detail in reference [4].

Rotating relativistic stars set local inertial frames into rotation with respect to the distant stars. An object falling from rest at great distance toward a rotating star would fall—not toward its center but would acquire

---

[5]The density at which quantum gravity would be relevant is $10^{78}$ higher than found in neutron stars.

[6]The limiting mass is also sometimes referred to as the Chandrasekhar mass, or the Oppenheimer–Volkoff limit.

an ever larger angular velocity as it approaches. The effect of rotating stars on the fabric of spacetime acts back upon the structure of the stars and so is a topic found in reference [4].

## 1.3   Compact Stars and Dense-Matter Physics

The physics of dense matter such as in neutron stars is not as simple as the final resting place of neutrons imagined by Baade and Zwicky. The constitution of matter at the high densities attained in a neutron star—the particle types, abundances and their interactions—pose challenging problems in nuclear and particle physics. How should matter at supernuclear densities be described? In addition to nucleons, what exotic baryon species constitute it? Does a transition in phase from quarks confined in nucleons to the deconfined phase of quark matter occur in the density range of such stars? And how is the transition to be calculated? What new structure is introduced into the star? Do other phases like pion or kaon condensates play a role in their constitution? We do not have definitive answers to these questions. They are of intense importance to theorists and experimentalists alike: the purpose of building large particle accelerators at CERN is aimed at answering them. But perhaps some of the exotic states of matter envisioned by theorists can exist only in neutron stars. In some cases there are subtle signals that may be detectable [5].

In Fig. 1.1 we show a computation of the possible constitution and interior crystalline structure of a neutron star near the limiting mass of such stars. Only now are we beginning to appreciate the complex and marvelous structure of these objects [7]. Surely the study of neutron stars and their astronomical realization in pulsars will serve as a guide in the search for a solution to some of the fundamental problems of dense many-body physics both at the level of nuclear physics—the physics of baryons and mesons— and ultimately at the level of their constituents—quarks and gluons. And neutron stars may be the only objects in which a Coulomb lattice structure (Fig. 1.1) formed from two phases of one and the same substance (hadronic matter) exists.

We do not know from experiment what the properties of superdense matter are. However we can be guided by certain general principles in our investigation of the possible forms that compact stars may take. Some of the possibilities lead to quite striking consequences that may in time be observable. The rate of discovery of new pulsars, X-ray neutron stars, and other high-energy phenomena associated with neutron stars is astonishing, and was unforeseen a dozen years ago.

White dwarfs are the cores of stars whose demise is less spectacular than a supernova—a more quiescent thermal expansion of the envelope of a low-mass star into a planetary nebula, and the subsidence of its central region

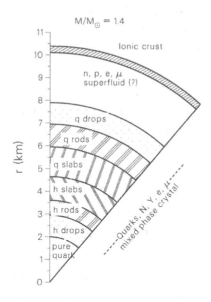

FIGURE 1.1. A section through a neutron star model that contains an inner sphere of pure quark matter surrounded by a crystalline region of mixed phase hadronic and quark matter. The mixed phase region consists of various geometrical objects of the rare phase immersed in the dominant one labeled by h(adronic) drops, h(adronic) rods, ..., immersed in quark matter, through to q(uark) drops ... immersed in hadronic matter. The particle composition of these regions is quarks, nucleons, hyperons, and leptons. A liquid of neutron star matter containing nucleons and leptons surrounds the mixed phase. A thin crust of heavy ions forms the stellar surface [6].

to form a white dwarf. (Planetary nebula is a misnomer inasmuch as it has nothing to do with a planet. Early astronomers interpreted the centers of such objects, which are white dwarfs, as planets.) The constituents of white dwarfs are nuclei immersed in an electron gas and therefore arranged in a Coulomb lattice because of the opposite electrical charge of gas and nuclei. White dwarfs are supported against collapse by the Fermi pressure of degenerate electrons—while neutron stars—are supported by the Fermi pressure of degenerate nucleons and by the repulsion of nucleons at close distance. White dwarfs pose less severe and less fundamental problems than neutron stars. The nuclei will comprise varying proportions of helium, carbon, and oxygen, and in some cases heavier elements like magnesium, depending on how far in the chain of exothermic nuclear fusion reactions the precursor star burned before it was disrupted by instabilities leaving behind the dwarf. White dwarfs are barely relativistic.

Of a vastly different nature than neutron stars are *strange stars*. Like neutron stars, strange stars, if they exist, are very dense, of the same order as neutron stars. However their very existence hinges on a hypothesis that at first sight seems absurd. According to the hypothesis, sometimes referred to as the *strange-matter hypothesis*, quark matter—consisting of an approximately equal number of up, down and strange quarks—has an equilibrium energy per nucleon that is lower than the mass of the nucleon or the energy per nucleon of the most bound nucleus, iron. In other words, under the hypothesis, strange quark matter is the *absolute* ground state of the strong interaction [8].

We customarily find that systems, if not in their ground state, readily decay to it. Of course this is not always so. Even in well known objects like nuclei, there are certain excited states whose structure is such that the transition to the ground state is hindered. The first excited state of $^{180}$Ta has a half-life of $10^{15}$ years, five orders of magnitude longer than the age of the universe! The strange-matter hypothesis is consistent with the present universe—a long-lived excited state—if strange matter is the ground state. The structure of strange stars is fascinating as are some of their properties. But that subject does not fall within the narrower aims of this book in laying down the General Theory of Relativity. The subject is treated in reference [4].

# 2

# Special Relativity

> "If I pursue a beam of light with velocity $c$, I should observe such a beam ...at rest. [ie. the oscillation would be frozen]. However, there appears to be no such thing ..."
> A. Einstein at age 16 [as recalled many years later]

> "The views of space and time which I wish to lay before you have sprung from the soil of experimental physics, and therein lies their strength. They are radical. Henceforth space by itself, and time by itself, are doomed to fade away into mere shadows, and only a kind of union of the two will preserve an independent reality."
> H. Minkowski [9].

The *principle of relativity* in physics goes back to Galileo who asserted that the laws of nature are the same in *all* uniformly moving laboratories. The relativity principle, stated in the narrow terms of reference frames in uniform motion, referred to as inertial frames, implies the existence of an absolute space relative to which they move. The notion of the absoluteness of time goes back to time immemorial. A *Galilean* transformation assumes the absoluteness of both space and time:

$$x' = x - a - vt, \quad y' = y, \quad z' = z, \quad t' = t - b. \qquad (2.1)$$

Newton's second law $F_x = m\, d^2x/dt^2$ is evidently invariant under this transformation if one assumes that force and mass are independent of the state of motion. However, in contrast to Galileo's principle, Einstein made the important distinction between reference frames that simply are in uniform motion, and those that are in uniform motion *with respect to each other*. This distinction makes all the difference in the world.

In contrast to Newton's equations, Maxwell's equations do not take on the same form if subjected to a Galilean transformation whereas under a *Lorentz transformation* they do.[1] This fact led Einstein to the postulate that the speed of light, and every law of nature, is the same in all inertial (unaccelerated) systems.

That is the historical role that light speed played in the discovery of Special Relativity, and the reason for the undoubted influence that the

---

[1]See, for example, Ref. [10] for the Lorentz invariant form of Maxwell's equations.

Michelson–Morley experiment [11] had on the early acceptance of the theory.

However, the underlying physics is quite different from how it appears in the historical development of the Special Theory. The speed of light need not have been postulated as an invariant.

H. Minkowski (1849-1909) realized soon after Einstein's epochal discovery in 1905 that the spacetime manifold of our world is not Euclidean space but is pseudo-Euclidean and we call it the Minkowski Manifold having such a nature that $d\tau^2 \equiv k^2 dt^2 - dx^2 - dy^2 - dz^2$ is invariant in the absence of gravity. The constant $k$ is a conversion factor between length and time.

Minkowski's fundamental discovery was inspired by Einstein's theory, which was based on the *postulate* of the constancy of light speed. However, as Minkowski realized, the constancy of light speed is a *consequence* of the nature of the spacetime manifold of our universe. In other words, Special Relativity is a consequence of the local spacetime manifold in which we live [12, 13].

That the constancy of the speed of light is a *property* of the local spacetime manifold is clearly illustrated by a thought experiment proposed by Swiatecki [12, 13]. He shows that the invariance of the differential interval between spacetime events (sometimes called the proper time, or the Minkowski invariant),

$$d\tau^2 = k^2 dt^2 - dx^2 - dy^2 - dz^2 \,, \tag{2.2}$$

can be verified (at least in principle) *without* resort to propagation of light signals, but with *only* measuring rods and clocks. And if it were technically feasible to perform the experiment with sufficient accuracy, $k$ would be measured and its value would be found to equal $c$ if indeed, the photon is massless, as we assume hereafter.

As an historical note, we add that Voigt observed in 1887 that $\Box\phi = 0$ preserved its form under a transformation that differed from the Lorentz transformation by only a scale factor [14]. In fact we will see shortly that the d'Alembertian $\Box$ is a Lorentz scalar. Consequently,

$$\left(\frac{1}{c^2}\frac{\partial^2}{\partial t^2} - \nabla^2\right)\phi = 0 \,, \tag{2.3}$$

informs us that a disturbance described by a wave equation for a massless particle in Minkowski spacetime propagates with velocity $c$ in vacuum as viewed from this and any other reference frame connected to it by a Lorentz transformation.

Having now stressed that it is by assumption of zero mass for the photon that we can use light velocity $c$ in the Minkowski invariant, we shall in the sequel make this assumption. Nevertheless, as a matter of principle, let us not loose sight of the fact that the invariance of the propertime is a property of the spacetime of the world we live in, and an empirical determination of the constant $k$ appearing in it would find its values to be $c$, within experimental error.

## 2.1  Lorentz Invariance

The Special Theory of Relativity, which holds in the absence of gravity, plays a central role in physics. Even in the strongest gravitational fields the laws of physics must conform to it in a sufficiently small locality of any spacetime event. That was a fundamental insight of Einstein. Consequently, the Special Theory plays a central role in the development of the General Theory of Relativity and its applications.

### 2.1.1  Lorentz Transformations

The Lorentz transformation leaves invariant the *proper time* or *differential interval* in Minkowski spacetime

$$d\tau^2 = dt^2 - dx^2 - dy^2 - dz^2 \quad \text{(units } c = 1) \tag{2.4}$$

as measured by observers in frames moving with *constant relative velocity* (called inertial frames because they move freely under the action of no forces). The Minkowski manifold also implies an absolute spacetime in which spacetime events that can be connected by a Lorentz transformation lie within the cone defined by $d\tau = 0$. "Absolute" means unaffected by any physical conditions. This was the same criticism that Einstein made of Newton's space *and* time, and the one that powered his search for a new theory in which the expression of physical laws does not depend on the frame of reference, but, nevertheless, in which Lorentz invariance would remain a *local* property of spacetime. ( *Local* in this context, means "in a sufficiently small region of spacetime".) We will develop General Relativity, which extends the relativity principle to arbitrary frames in a gravity-endowed universe, not just un-accelerated frames in relative uniform motion. Here we review briefly the Special Theory.

A *pure* Lorentz transformation is one without spatial rotation, while a general Lorentz transformation is the product of a rotation in space and a pure Lorentz transformation. The pure Lorentz transformation is sometimes also referred to as a *boost*. For convenience, define

$$x^0 = t, \quad x^1 = x, \quad x^2 = y, \quad x^3 = z. \tag{2.5}$$

(In spacetime a point, such as that above, is sometimes referred to as an *event*.) The linear homogeneous transformation connecting two reference frames can be written

$$x'^\mu = \Lambda^\mu_{\ \nu} x^\nu. \tag{2.6}$$

(We shall use the convenient notation introduced by Einstein whereby repeated indices are summed—Greek over time and space, Roman over space alone.)

Any set of four quantities $A^\mu$ ($\mu = 0, 1, 2, 3$) that transforms under a change of reference frame in the same way as the coordinates is called a *contravariant* Lorentz four-vector,

$$A'^\mu = \Lambda^\mu{}_\nu A^\nu . \tag{2.7}$$

The invariant interval (also variously called the proper time, the line element, or the separation formula) can be written

$$d\tau^2 = \eta_{\mu\nu}\, dx^\mu\, dx^\nu , \tag{2.8}$$

where $\eta_{\mu\nu}$ is the Minkowski metric which, in rectilinear coordinates, is

$$\eta_{\mu\nu} \equiv \begin{pmatrix} 1 & 0 & 0 & 0 \\ 0 & -1 & 0 & 0 \\ 0 & 0 & -1 & 0 \\ 0 & 0 & 0 & -1 \end{pmatrix} . \tag{2.9}$$

The condition of the invariance of $d\tau^2$ is

$$\eta_{\alpha\beta}\, dx^\alpha\, dx^\beta = \eta_{\mu\nu}\, dx'^\mu\, dx'^\nu = \eta_{\mu\nu}\Lambda^\mu{}_\alpha\Lambda^\nu{}_\beta\, dx^\alpha\, dx^\beta . \tag{2.10}$$

Since this holds for any $dx^\alpha, dx^\beta$ we conclude that the $\Lambda^\mu{}_\nu$ must satisfy the fundamental relationship assuring invariance of the proper time:

$$\eta_{\alpha\beta} = \eta_{\mu\nu}\Lambda^\mu{}_\alpha\Lambda^\nu{}_\beta . \tag{2.11}$$

Transformations that leave $d\tau^2$ invariant also leave the speed of light the same in all inertial systems; this is so because if $d\tau = 0$ in one system, it is true in all, and the content of $d\tau = 0$ is that $d\mathbf{x}/dt = 1$ (with units $c = 1$).

Let us find the transformation matrix $\Lambda^\mu{}_\alpha$ for the special case of a boost along the $x$-axis. In this case it is clear that

$$x'^2 = x^2, \quad x'^3 = x^3, \tag{2.12}$$

and, moreover, that $x'^0$ and $x'^1$ cannot involve $x^2$ and $x^3$. So,

$$\begin{aligned} x'^0 &= \Lambda^0{}_0 x^0 + \Lambda^0{}_1 x^1 \\ x'^1 &= \Lambda^1{}_0 x^0 + \Lambda^1{}_1 x^1 , \end{aligned} \tag{2.13}$$

with the remaining $\Lambda$ elements zero. So the above quadratic form in $\Lambda$ yields the three equations,

$$\begin{aligned} 1 &= (\Lambda^0{}_0)^2 - (\Lambda^1{}_0)^2 \\ -1 &= (\Lambda^0{}_1)^2 - (\Lambda^1{}_1)^2 \\ 0 &= \Lambda^0{}_0\Lambda^0{}_1 - \Lambda^1{}_0\Lambda^1{}_1 . \end{aligned} \tag{2.14}$$

To get a fourth equation, suppose that the origins of the two frames in uniform motion coincide at $t = 0$ and the primed x-axis, $x'^1$, is moving along $x^1$ with velocity $v$. That is, $x^1 = vt$ is the equation of the *primed* origin as it moves along the *unprimed* x-axis. The equation for the primed coordinate is

$$0 = x'^1 = \Lambda^1{}_0 x^0 + \Lambda^1{}_1 x^1 = (\Lambda^1{}_0 + \Lambda^1{}_1 v)t \tag{2.15}$$

or

$$\Lambda^1{}_0 = -\Lambda^1{}_1 v. \tag{2.16}$$

The four equations can now be solved with the result,

$$\begin{aligned} \Lambda^0{}_0 &= \Lambda^1{}_1 = \gamma \\ \Lambda^1{}_0 &= \Lambda^0{}_1 = -v\gamma \\ \Lambda^2{}_2 &= \Lambda^3{}_3 = 1, \end{aligned} \tag{2.17}$$

where

$$\gamma \equiv (1 - v^2)^{-1/2} \equiv \cosh\theta, \quad v\gamma \equiv \sinh\theta, \quad v \equiv \tanh\theta. \tag{2.18}$$

So

$$\begin{aligned} x'^0 &= x^0 \cosh\theta - x^1 \sinh\theta \\ x'^1 &= -x^0 \sinh\theta + x^1 \cosh\theta \\ x'^2 &= x^2, x'^3 = x^3. \end{aligned} \tag{2.19}$$

The combination of two boosts in the same direction, say $v_1$ and $v_2$, corresponds to $\theta = \theta_1 + \theta_2$. A boost in an arbitrary direction with the *primed* axis having velocity $\mathbf{v} = (v^1, v^2, v^3)$ relative to the *unprimed* is

$$\begin{aligned} \Lambda^0{}_0 &= \gamma \\ \Lambda^0{}_j &= \Lambda^j{}_0 = -v^j\gamma \\ \Lambda^j{}_k &= \Lambda^k{}_j = \delta^j_k + (\gamma - 1)v^j v^k / \mathbf{v}^2. \end{aligned} \tag{2.20}$$

For a spatial rotation, say in the x-y plane, the transformation for a positive rotation about the common z-axis is

$$\begin{aligned} x'^1 &= x^1 \cos\omega + x^2 \sin\omega \\ x'^2 &= -x^1 \sin\omega + x^2 \cos\omega \\ x'^0 &= x^0, \quad x'^3 = x^3. \end{aligned} \tag{2.21}$$

Transformation of vectors according to either of the above, or a product of them, preserves the invariance of the interval $d\tau^2$. For convenience they can be written in matrix form as

$$\Lambda \equiv \begin{pmatrix} \cosh\theta & -\sinh\theta & 0 & 0 \\ -\sinh\theta & \cosh\theta & 0 & 0 \\ 0 & 0 & 1 & 0 \\ 0 & 0 & 0 & 1 \end{pmatrix} \tag{2.22}$$

$$R \equiv \begin{pmatrix} 1 & 0 & 0 & 0 \\ 0 & \cos\omega & \sin\omega & 0 \\ 0 & -\sin\omega & \cos\omega & 0 \\ 0 & 0 & 0 & 1 \end{pmatrix}. \qquad (2.23)$$

### 2.1.2   TIME DILATION

Let there be two identical clocks at rest at $x = 0$. The clocks tick at equal intervals $dt$. The propertime in this frame is

$$d\tau = \sqrt{dt^2 - dx^2} = dt. \qquad (2.24)$$

An observer takes one of the clocks and moves away along the $x$-axis with velocity $v$. He sees the first clock at $dx' = -v\,dt$. The propertime expressed in his frame is

$$d\tau' = \sqrt{dt'^2 - dx'^2} = dt'\sqrt{1 - v^2} \qquad (2.25)$$

But the two events are connected by a Lorentz transformation and so their propertimes are the same. Consequently

$$dt' = dt/\sqrt{1 - v^2}. \qquad (2.26)$$

This expresses the dilation of time as measured by observers in uniform relative motion. The converse holds for length measurements.

### 2.1.3   COVARIANT VECTORS

Two contravariant Lorentz vectors such as

$$A^\mu \equiv (A^0, A^1, A^2, A^3) \qquad (2.27)$$

and $B^\mu$ may be used to create a *scalar* product (Lorentz scalar)

$$A' \cdot B' \equiv \eta_{\mu\nu} A'^\mu B'^\nu = \eta_{\mu\nu} \Lambda^\mu{}_\alpha \Lambda^\nu{}_\beta A^\alpha B^\beta = \eta_{\alpha\beta} A^\alpha B^\beta \equiv A \cdot B. \quad (2.28)$$

Because of the minus signs in the Minkowski metric we have

$$A \cdot B = A^0 B^0 - \mathbf{A} \cdot \mathbf{B}, \qquad (2.29)$$

and the *covariant* Lorentz vector is defined by

$$A_\mu \equiv (A^0, -A^1, -A^2, -A^3). \qquad (2.30)$$

A *covariant* Lorentz vector is obtained from its contravariant dual by the process of lowering indices with the metric tensor,

$$A_\mu = \eta_{\mu\nu} A^\nu. \qquad (2.31)$$

Conversely, raising of indices is achieved by

$$A^\mu = \eta^{\mu\nu} A_\nu \,.\tag{2.32}$$

It is straightforward to show that

$$\eta^{\mu\alpha} \eta_{\alpha\nu} \equiv \eta^\mu_\nu = \delta^\mu_\nu \,,\tag{2.33}$$

where

$$\delta^\mu_\nu = \begin{cases} 1 & \text{if } \mu = \nu \\ 0 & \text{otherwise} \end{cases}\tag{2.34}$$

is the Kroneker delta. It follows that

$$\eta_{\mu\nu} = \eta^{\mu\nu} \,.\tag{2.35}$$

The Lorentz transformation for a covariant vector is written in analogy with that of a contravariant vector:

$$A'_\mu = \Lambda_\mu{}^\nu A_\nu \,.\tag{2.36}$$

To obtain the elements of $\Lambda_\mu{}^\nu$ we write the above in two different ways,

$$\eta_{\mu\beta} \Lambda^\beta{}_\alpha A^\alpha = \eta_{\mu\beta} A'^\beta = A'_\mu = \Lambda_\mu{}^\nu A_\nu = \Lambda_\mu{}^\nu \eta_{\nu\alpha} A^\alpha \,.\tag{2.37}$$

This holds for arbitrary $A^\mu$ so

$$\Lambda_\mu{}^\nu = \eta_{\mu\alpha} \Lambda^\alpha{}_\beta \eta^{\beta\nu} \,.\tag{2.38}$$

Using (2.33) in the above we get the inverse relationship

$$\Lambda^\mu{}_\nu = \eta^{\mu\alpha} \Lambda_\alpha{}^\beta \eta_{\beta\nu} \,.\tag{2.39}$$

Multiplying (2.38) by $A^\mu{}_\sigma$, summing on $\mu$, and employing the fundamental condition of invariance of the proper time (2.11) we find

$$\Lambda^\mu{}_\sigma \Lambda_\mu{}^\tau = \delta^\tau_\sigma \,.\tag{2.40}$$

We can now invert (2.6) and find that $\Lambda_\mu{}^\nu$ is the inverse Lorentz transformation,

$$x^\mu = \Lambda_\nu{}^\mu x'^\nu \,.\tag{2.41}$$

The elements of the inverse transformation are given in terms of (2.17) or (2.20) by (2.38). We have

$$\begin{aligned} \Lambda_0{}^0 &= \Lambda_1{}^1 = \gamma \,, \\ \Lambda_1{}^0 &= \Lambda_0{}^1 = v\gamma \,, \\ \Lambda_2{}^2 &= \Lambda_3{}^3 = 1 \,. \end{aligned}\tag{2.42}$$

A boost in an arbitrary direction with the primed axis having velocity $\mathbf{v} = (v^1, v^2, v^3)$ relative to the unprimed is

$$
\begin{aligned}
\Lambda_0^{\;0} &= \gamma, \\
\Lambda_0^{\;j} &= \Lambda_j^{\;0} = v^j \gamma, \\
\Lambda_j^{\;k} &= \Lambda_k^{\;j} = \delta_k^j + (\gamma - 1) v^j v^k / \mathbf{v}^2.
\end{aligned}
\tag{2.43}
$$

The *velocity* is a vector of particular interest and defined as

$$
u^\mu = \frac{dx^\mu}{d\tau}.
\tag{2.44}
$$

Because $d\tau$ is an invariant scalar and $dx^\mu$ is a vector, $u^\mu$ is obviously a contravariant vector. From the expression for the invariant interval we have

$$
d\tau = \sqrt{1 - \mathbf{v}^2}\, dt, \qquad \mathbf{v} = \frac{d\mathbf{r}}{dt}
\tag{2.45}
$$

with $\mathbf{r} = (x^1, x^2, x^3)$; it therefore follows that

$$
u^0 \equiv \frac{dt}{d\tau} = \gamma, \quad u^i \equiv \frac{dx^i}{d\tau} = \frac{dx^i}{dt}\frac{dt}{d\tau} = v^i \gamma,
\tag{2.46}
$$

or

$$
\begin{aligned}
u^\mu &= \gamma(1, v^1, v^2, v^3), \quad u_\mu = \gamma(1, -v^1, -v^2, -v^3), \\
u^\mu u_\mu &= 1.
\end{aligned}
\tag{2.47}
$$

The transformation of a tensor under a Lorentz transformation follows from (2.7) and (2.36) according to the position of the indices; for example,

$$
T'^{\mu\nu} = \Lambda^\mu_{\;\alpha} \Lambda^\nu_{\;\beta} T^{\alpha\beta}.
\tag{2.48}
$$

We note that according to (2.11), the Minkowski metric $\eta^{\mu\nu}$ is a tensor; moreover, it has the same constant form in every Lorentz frame.

## 2.1.4   ENERGY AND MOMENTUM

The relativistic analogue of Newton's law $F = ma$ is

$$
F^\mu = m \frac{d^2 x^\mu}{d\tau}
\tag{2.49}
$$

and the four-momentum is

$$
p^\mu = m \frac{dx^\mu}{d\tau}.
\tag{2.50}
$$

Hence, from (2.44) and (2.45)

$$
p^0 \equiv E = m\gamma
$$
$$
\mathbf{p} = m\gamma \mathbf{v}.
\tag{2.51}
$$

## 2.1.5 ENERGY-MOMENTUM TENSOR OF A PERFECT FLUID

A perfect fluid is a medium in which the pressure is isotropic in the rest frame of each fluid element, and shear stresses and heat transport are absent. If at a certain point the velocity of the fluid is **v**, an observer with this velocity will observe the fluid in the neighborhood as isotropic with an energy density $\epsilon$ and pressure $p$. In this local frame the energy-momentum tensor is

$$T'^{\mu\nu} \equiv \begin{pmatrix} \epsilon & 0 & 0 & 0 \\ 0 & p & 0 & 0 \\ 0 & 0 & p & 0 \\ 0 & 0 & 0 & p \end{pmatrix}. \tag{2.52}$$

As viewed from an arbitrary frame, say the laboratory system, let this fluid element be observed to have velocity **v**. According to (2.41) we obtain the transformation

$$T^{\mu\nu} = \Lambda_\alpha^{\ \mu} \Lambda_\beta^{\ \nu} T'^{\alpha\beta} . \tag{2.53}$$

The elements of the transformation are given by (2.42) in the case that the fluid element is moving with velocity $v$ along the laboratory x-axis, or by (2.43) if it has the general velocity **v**. It is easy to check that in the arbitrary frame

$$T^{\mu\nu} = -p\eta^{\mu\nu} + (p + \epsilon)u^\mu u^\nu \tag{2.54}$$

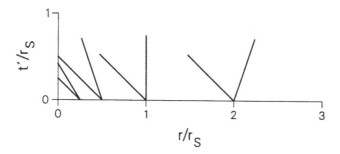

FIGURE 2.1. Future light cones at radial distances both inside and outside a black hole Schwarzschild radius $r_s \equiv 2GM/c^2$. The possible futures of any event at the vertex of each cone, lies *within* the cone. Light propagates along the cone itself. On the scale of distance relative to the Schwarzschild radius of the black hole, the cones narrow and are tipped toward the black hole. At the critical radius, the outer edge of the cone is vertical; not even light can escape. Within the black hole, light can propagate only inward, as with anything else. Future points in spacetime of a *material particle* at the vertex of a cone can lie only *within* the cone.

and reduces to the diagonal form above when $\mathbf{v} = \mathbf{0}$. We have used the four-velocity $u^{\mu}$ defined above by (2.46). Relative to the laboratory frame it is the four-velocity of the fluid element.

### 2.1.6  LIGHT CONE

The vanishing of the proper time interval, $d\tau = 0$, given by (2.4) defines a cone in the four-dimensional space $x^{\mu}$ with the time axis as the axis of the cone. Events separated from the vertex event for which the proper time, (or invariant interval) vanishes ($d\tau = 0$), are said to have null separation. They can be connected to the event at the vertex by a light signal. Events separated from the vertex by a real interval $d\tau^2 > 0$ can be connected by a subluminal signal—a material particle can travel from one event to the other. An event for which $d\tau^2 < 0$ refers to an event outside the two cones; a light signal cannot join the vertex event to such an event. Therefore, events in the cone with $t$ greater than that of the vertex of the cone lie in the future of the event at the vertex, while events in the other cone lie in its past. Events lying outside the cone are not causally connected to the vertex event.

# 3

# General Relativity

"Scarcely anyone who fully comprehends this theory can escape its magic."
*A. Einstein*

"Beauty is truth, truth beauty–that is all
Ye know on earth, and all ye need to know."
*J. Keats*     Ode on a Grecian Urn

General Relativity—Einstein's theory of gravity—is the most beautiful and elegant of physical theories. It is the foundation of cosmology—the subject that traces the evolution of the universe from its first intensely hot and dense beginning to its possible futures.[1] General Relativity is also the foundation for our understanding of compact stars. Neutron stars and black holes Can be understood correctly only in General Relativity as formulated by Einstein [15, 16]. Dense objects like neutron stars could also exist in Newton's theory, but they would be very different objects. Chandrasekhar found (in connection with white dwarfs) that all degenerate stars have a maximum possible mass [17, 18]. In Newton's theory such a maximum mass is attained only asymptotically when all Fermions, whose pressure supports the star, are ultra relativistic. Under such conditions, stars populated by the three heavy quarks—known as charm, truth, and beauty—would exist. However, such stars do not occur in Einstein's theory because the maximum-possible-mass star is not sufficiently dense, even at its center; therefore they cannot exist in nature.

Einstein sought the answer to a question that must have seemed to his contemporaries, and all who preceded him, as of no consequence: *What meaning is attached to the absolute equality of inertial and gravitational masses?* If all bodies move in gravitational fields in precisely the same way, no matter what their constitution or binding forces, then this means that their motion has nothing to do with their nature, but rather with the na-

---

[1]Einstein, himself, never did apply his theory to the evolution of the cosmos. Indeed, when he discovered the theory (1915), it was a canon of western thought that the world lasted from "everlasting to everlasting." Edwin Hubble's 1927 discovery of universal expansion shook that belief.

*ture of spacetime*. And if spacetime determines the motion of bodies, then according to the notion of action and reaction, this implies that spacetime in turn is *shaped by bodies and their motion*.

Beautiful or not, the predictions of theory have to be tested. The first three tests of General Relativity were proposed by Einstein, the gravitational redshift, the deflection of light by massive bodies and the perihelion shift of Mercury. The latter had already been measured. Einstein computed the anomalous part of the precession to be 43 arcseconds per century, in agreement with the measurement of $42.98 \pm 0.04$. A fourth test was suggested by Shapiro in 1964—the time delay in the radar echo of a signal sent to a planet whose orbit is carrying it toward superior conjunction[2] with the Sun. Eventually agreement to 0.1 percent with the prediction of Einstein's theory was achieved in these difficult and remarkable experiments. It should be noted that all of the above tests involved weak gravitational fields.

In testing General Relativity, the crowning achievement was the cumulative 20-year study by Taylor, Hulse and their colleagues of the Hulse–Taylor pulsar binary that was discovered in 1974. Their work yielded a measurement of 4.22663 degrees *per year* for the periastron shift of the orbit of the neutron star binary and a measurement of the decay of the orbital period by $7.60 \pm 0.03 \times 10^{-7}$ seconds per year. This rate of decay agrees to less than 1% with careful calculations of the effect of energy loss through gravitational radiation as predicted by Einstein's theory [19]. A fuller discussion of these experiments and other intricacies involved in the tests of relativity can be found in the book by Will [20]. Since these early experiments, more accurate tests are being made by Dick Manchester and collaborators at Parkes Observatory in Australia, who have discovered a tighter orbiting pair of pulsars (neutron stars) — "We have verified GR to 0.1% already in two years" — ten times better than the early experiment." (Private communication: R. N. Manchester, 6/15/2005).

The goal of this chapter is to provide a rigorous derivation of the Oppenheimer–Volkoff equations that describe the structure of relativistic stars. For this purpose we need to understand General Relativity.

## 3.1   Scalars, Vectors, and Tensors in Curvilinear Coordinates

In the last chapter we dealt with inertial frames of reference in flat spacetime. We now wish to allow for curvilinear coordinates. Our scalars, vectors, and tensors now refer to a point in spacetime. Their components refer to the reference frame at that point.

---

[2]Superior conjunction refers to the situation when the Earth and the planet are on opposite sides of the Sun.

A scalar field $S(x)$ is a function of position, but its value does not depend on the coordinate system. An example is the temperature as registered on thermometers located in various rooms in a house. Each registered temperature may be different, and therefore is a function of position, but independent of the coordinates used to specify the locations:

$$S'(x') = S(x). \tag{3.1}$$

A vector is a quantity whose components change under a coordinate transformation. One important vector is the displacement vector between adjacent points. Near the point $x^\mu$ we consider another, $x^\mu + dx^\mu$. The four displacements $dx^\mu$ are the components of a vector. Choose units so that time and distance are measured in the same units ($c = 1$). In Cartesian coordinates we can write the *invariant interval* $d\tau$ of the Special Theory of Relativity, sometimes called the *proper time*, as

$$d\tau^2 = (dx^0)^2 - (dx^1)^2 - (dx^2)^2 - (dx^3)^2. \tag{3.2}$$

Under a coordinate transformation from these rectilinear coordinates to arbitrary coordinates, $x^\mu \to x'^\mu$, we have (from the rules of partial differentiation)

$$dx'^\mu = \frac{\partial x'^\mu}{\partial x^\nu} dx^\nu. \tag{3.3}$$

As always, repeated indices are summed. We can also write the inverse of the above equation and substitute for the spacetime differentials in the invariant interval to obtain an equation of the form

$$d\tau^2 = g_{\mu\nu} dx^\mu dx^\nu, \tag{3.4}$$

where the $g_{\mu\nu}$ are defined in terms of products of the partial derivatives of the coordinate transformation.

Depending on the nature of the coordinate system, say rectilinear, oblique, or curvilinear, or on the presence of a gravitational field, the invariant interval may involve bilinear products of different $dx^\mu$, and the $g_{\mu\nu}$ will be functions of position and time. The $g_{\mu\nu}$ are field quantities—the components of a tensor called the *metric tensor*. Because the $g_{\mu\nu}$ appear in a quadratic form (3.4), we may take them to be symmetric:

$$g_{\mu\nu} = g_{\nu\mu}. \tag{3.5}$$

In regions of spacetime for which the rectilinear system of the Special Theory of Relativity holds, the metric tensor $g_{\mu\nu}$ is equal to the Minkowski tensor (2.9). In fact, as we shall see, Special Relativity holds *locally* anywhere at any time. We shall refer to reference frames in which the metric is given by the Minkowski tensor as *Lorentz frames*.

The invariant interval or proper time $d\tau$ is real for a timelike interval and imaginary for a spacelike interval.[3] The notation *proper time* is seen to be appropriate because, when two events occur at the same *space* point, what remains of the invariant interval is $dt$.

Any four quantities $A^\mu$ that transform as $dx^\mu$ comprise a *contravariant vector*

$$A'^\mu = \frac{\partial x'^\mu}{\partial x^\nu} A^\nu ,  \tag{3.6}$$

and

$$g_{\mu\nu} A^\mu A^\nu \equiv A^2  \tag{3.7}$$

is its invariant squared length. It is obviously invariant under the same transformations that leave (3.2) invariant because the four quantities $A^\mu$ form a four-vector like $dx^\mu$.

A *covariant* vector can be obtained through the process of *lowering* indices with the metric tensor:

$$A_\mu = g_{\mu\nu} A^\nu .  \tag{3.8}$$

In terms of this vector, the magnitude equation (3.7) can be written as

$$A_\mu A^\mu = A^2 .  \tag{3.9}$$

Let $A^\mu$ and $B^\mu$ be distinct contravariant vectors. Then $A^\mu + \lambda B^\mu$ is also a contravariant vector for all finite $\lambda$. The quantity

$$g_{\mu\nu} \left( A^\mu + \lambda B^\mu \right) \left( A^\nu + \lambda B^\nu \right)$$

is the invariant squared length. Because this is true for all $\lambda$, the coefficient of each power of $\lambda$ is also an invariant; for the linear term we find

$$g_{\mu\nu} \left( A^\mu B^\nu + B^\mu A^\nu \right) = 2 g_{\mu\nu} A^\mu B^\nu ,  \tag{3.10}$$

where we have used the symmetry of $g_{\mu\nu}$. Thus, we obtain the invariant scalar product of two vectors:

$$A \cdot B \equiv g_{\mu\nu} A^\mu B^\nu = A_\nu B^\nu .  \tag{3.11}$$

To derive the transformation law for a covariant vector use the fact, just proven, that $A_\mu B^\mu$ is a scalar. Then using the law of transformation of a contravariant vector (3.6), we have

$$A'_\mu B'^\mu = A_\alpha B^\alpha = A_\alpha \frac{\partial x^\alpha}{\partial x'^\mu} B'^\mu ,  \tag{3.12}$$

---

[3]The opposite convention $ds^2 = -d\tau^2$ could also be employed. The interval $ds$ is often referred to as the line element.

where $A'_\mu$ is the same vector as $A_\mu$ but referred to the primed reference frame. From the above equation it follows that

$$\left(A'_\mu - \frac{\partial x^\alpha}{\partial x'^\mu} A_\alpha\right) B'^\mu = 0 . \tag{3.13}$$

Because $B^\mu$ is any vector, the quantity in brackets must vanish; thus we have the law of transformation of a covariant vector,

$$A'_\mu = \frac{\partial x^\nu}{\partial x'^\mu} A_\nu . \tag{3.14}$$

Compare this transformation law with that of (3.6).

Let the determinant of $g_{\mu\nu}$ be $g$,

$$g = \det|g_{\mu\nu}| . \tag{3.15}$$

As long as $g$ does not vanish, the equations (3.8) can be inverted. Let the coefficients of the inverse be called $g^{\mu\nu}$. Then find

$$A^\nu = g^{\nu\mu} A_\mu . \tag{3.16}$$

Multiply (3.8) by $g^{\alpha\mu}$ and sum on $\mu$ with the result

$$A^\alpha = g^{\alpha\mu} A_\mu = g^{\alpha\mu} g_{\mu\nu} A^\nu , \tag{3.17}$$

or

$$(g^{\alpha\mu} g_{\mu\nu} - \delta^\alpha_\nu) A^\nu = 0 , \tag{3.18}$$

where $\delta^\alpha_\nu$ is the Kroneker delta. Because this equation holds for any vector, we have

$$g^\alpha_\beta \equiv g^{\alpha\mu} g_{\mu\beta} = \delta^\alpha_\beta . \tag{3.19}$$

The two $g$'s, one with subscripts, the other with superscripts, are inverses. In the same way as $g_{\mu\nu}$ can be used to lower an index, $g^{\mu\nu}$ can be used to raise one as in,

$$A^\mu = g^{\mu\nu} A_\nu . \tag{3.20}$$

Both $g$'s are symmetric;

$$g_{\mu\nu} = g_{\nu\mu}, \quad g^{\mu\nu} = g^{\nu\mu} . \tag{3.21}$$

The derivative of a scalar field $S(x) = S'(x')$ with respect to the components of a contravariant position vector yields a covariant vector field and, vice versa:

$$\frac{\partial S}{\partial x'^\mu} = \frac{\partial x^\nu}{\partial x'^\mu} \frac{\partial S}{\partial x^\nu} . \tag{3.22}$$

Accordingly, we shall sometimes use the abbreviations,

$$\partial_\mu \equiv \frac{\partial}{\partial x^\mu} \quad \text{and} \quad \partial^\mu \equiv \frac{\partial}{\partial x_\mu}, \tag{3.23}$$

especially in writing Lagrangians of fields. In relativity it is also useful to have an even more compact notation for the coordinate derivative—the "comma subscript":

$$S_{,\mu} \equiv \frac{\partial S}{\partial x^\mu}. \tag{3.24}$$

The d'Alembertian,

$$\Box = \partial_\mu \partial^\mu, \tag{3.25}$$

is manifestly a scalar.

Tensors are similar to vectors, but with more than one index. A simple tensor is one formed from the product of the components of two vectors, $A^\mu B^\nu$. But this is special because of the relationships between its components. A general tensor of the second rank can be formed by a sum of such products:

$$T^{\mu\nu} = A^\mu B^\nu + C^\mu D^\nu + \cdots. \tag{3.26}$$

The superscripts can be lowered as with a vector, either one index, or both,

$$T_\mu{}^\nu = g_{\mu\alpha} T^{\alpha\nu}, \quad T^\nu{}_\mu = T^{\nu\alpha} g_{\alpha\mu}, \quad T_{\mu\nu} = g_{\mu\alpha} g_{\nu\beta} T^{\alpha\beta}. \tag{3.27}$$

Similarly, we may have tensors of higher rank, either contravariant with respect to all indices, or covariant, or mixed. The position of the indices on the mixed tensor (the lower to the left or right of the upper) refers to the position of the index that was lowered. If $T^{\mu\nu}$ is symmetric, then $T^\mu{}_\nu = T_\nu{}^\mu$ and it is unimportant to keep track of the position of the index that has been lowered (or raised). But if $T^{\mu\nu}$ is antisymmetric, then the two orderings differ by a sign.

If two of the indices on a tensor, one a superscript the other a subscript, are set equal and summed, the rank is reduced by two. This process is called *contraction*. If it is done on a second-rank mixed tensor, the result is a scalar,

$$S = T^\mu{}_\mu = T_\mu{}^\mu. \tag{3.28}$$

When $T^{\mu\nu}$ is antisymmetric, the contractions $T_\mu{}^\mu$ and $T^\mu{}_\mu$ are identically zero.

The test of tensor character is whether the object in question transforms under a coordinate transformation in the obvious generalization of a vector. For example,

$$T'^\mu{}_\nu = \frac{\partial x'^\mu}{\partial x^\alpha} \frac{\partial x^\beta}{\partial x'^\nu} T^\alpha{}_\beta \tag{3.29}$$

is a tensor.

In general, we deal with curved spacetime in General Relativity. We must therefore deal with curvilinear coordinates. Vectors and tensors at a point in such a spacetime have components referring to the axis at that point. The components will change according to the above laws, depending on the way the axes change at that point. Therefore, the metric tensors $g_{\mu\nu}, g^{\mu\nu}$ cannot be constants. They are field quantities which vary from point to point. As we shall see, they can be referred to collectively as the gravitational field. Because the formalism of this section is expressed by local equations, it holds in curved spacetime, for curved spacetime is flat in a sufficiently small locality.

Because the derivative of a scalar field is a vector (3.22), one might have thought that the derivative of a vector field is a tensor. However, by checking the transformation properties one finds that this supposition is not true.

We have referred invariably to the $g_{\mu\nu}$ as tensors. Now we show that this is so. Let $A^\mu, B^\nu$ be arbitrary vector fields, and consider two coordinate systems such that the same point P has the coordinates $x^\mu$ and $x'^\mu$ when referred to the two systems, respectively. Then we have

$$g'_{\alpha\beta} A'^\alpha A'^\beta = g_{\mu\nu} A^\mu A^\nu = g_{\mu\nu} \frac{\partial x^\mu}{\partial x'^\alpha} \frac{\partial x^\nu}{\partial x'^\beta} A'^\alpha A'^\beta. \qquad (3.30)$$

Because this holds for arbitrary vectors, we find

$$g'_{\alpha\beta} = g_{\mu\nu} \frac{\partial x^\mu}{\partial x'^\alpha} \frac{\partial x^\nu}{\partial x'^\beta}, \qquad (3.31)$$

which, by comparison with (3.14), shows that $g_{\mu\nu}$ is a covariant tensor. Similarly $g^{\mu\nu}$ is a contravariant tensor:

$$g'^{\alpha\beta} = g^{\mu\nu} \frac{\partial x'^\alpha}{\partial x^\mu} \frac{\partial x'^\beta}{\partial x^\nu}. \qquad (3.32)$$

These are called the *fundamental tensors*. Of course, the above tensor character of the metric is precisely what is required to make the square of the interval $d\tau^2$ of (3.4) an invariant, as is trivially verified.

Mixed tensors of arbitrary rank transform, for each index, according to the transformation laws (3.6, 3.14) depending on whether the index is a superscript or a subscript, as can be derived in obvious analogy to the above manipulations.

Tensors and tensor algebra are very powerful techniques for carrying the consequences discovered in one frame to another. That the linear combination of tensors of the same rank and arrangement of upper and lower indices is also a tensor; that the direct product of two tensors of the same or different rank and arrangement of indices, $A^{\mu\cdots}_{\alpha\cdots} B^{\nu\cdots}_{\cdots} = T^{\mu\cdots\nu\cdots}_{\alpha\cdots}$ is also a tensor; and that contraction (defined above) of a pair of indices, one upper, one lower, produces a tensor of rank reduced by two—are all easy theorems that we do not need to prove, but only note in passing. Of particular note, if the difference of two tensors of the same transformation rule vanishes in one frame, then it vanishes in all (i.e., the two tensors are equal in all frames).

> "The possibility of explaining the numerical equality of inertia and gravitation by the unity of their nature gives to the general theory of relativity, according to my conviction, such a superiority over the conceptions of classical mechanics, that all the difficulties encountered in development must be considered as small in comparison." A. Einstein [16]

Eötvös established that all bodies have the same ratio of inertial to gravitational mass with high precision [21]. With an appropriate choice of units, the two masses are equal for all bodies to the accuracy established for the ratio. One might have expected such conceptually different properties, one having to do with inertia to motion $(m_I)$, the other with "charge" $(m_G)$, in an expression of mutual attraction between bodies, to be entirely different. The relation between the force exerted by the gravitational attraction of a body of mass $M$ at a distance $R$ upon the object, and the acceleration imparted to it are expressed by Newton's equation, valid for weak fields and small material velocities:

$$F_I = m_I a = G\frac{m_G M}{R^2} = F_G \,. \qquad (3.33)$$

Einstein reasoned that the near equality of two such different properties must be more than mere coincidence and that inertial and gravitational masses must be *exactly* equal:

$$m_I = m_G = m \,. \qquad (3.34)$$

The mass drops out! In that case all bodies experience precisely the same acceleration in a gravitational field, $a = m/R^2$, where $m$ is the mass of the accelerating body. A common acceleration due to gravity was presaged by Galileo's careful experiments performed centuries earlier. For all other forces that we know, the acceleration $(a = F/m)$ is inverse to the mass.

The equivalence of inertial and gravitational mass is established to high accuracy for atomic and nuclear binding energies.[4] Moreover, as a result of very careful lunar laser-ranging experiments, the earth and moon are found to fall with equal acceleration toward the Sun to a precision of almost 1 part in $10^{13}$, better than the most accurate Eötvös-type experiments on laboratory bodies. This exceedingly important test involving bodies of different gravitational binding was conceived by Nordtvedt [22]. The essentially null result establishes the so-called *strong* statement of equivalence

---

[4]Eötvös' experiments on such diverse media as wood, platinum, copper, glass, and other materials involve different molecular, atomic, and nuclear binding energies and different ratios of neutrons and protons.

of inertial and gravitational mass: Free bodies—no matter their nature or constituents, nor how much or little those constituents are bound, nor by what force—all move in the spacetime of an arbitrary gravitational field as if they were identical test particles! *Because their motion in spacetime has nothing to do with their nature, it evidently has to do with the nature of spacetime.*

Einstein felt certain that a deep meaning was attached to the equivalence; "The experimentally known matter independence of the acceleration of fall is .... a powerful argument for the fact that the relativity postulate has to be extended to coordinate systems which, relative to each other, are in non-uniform motion" [23]. This conviction led him to the formulation of the *equivalence principle*. The equivalence principle provides the link between the physical laws as we discern them in our laboratories and their form under any circumstance in the universe—more precisely, in arbitrarily strong and varying gravitational fields. It also provides a tool for the development of the theory of gravitation itself, as we shall see in the sequel.

The universe is populated by massive objects moving relative to one another. The gravitational field may be arbitrarily changing in time and space. However, the presence of gravity cannot be detected in a sufficiently small reference frame falling freely with a particle under no influence other than gravity. The particle will remain at rest in such a frame. It is a *local inertial frame*. A local inertial frame and a *local Lorentz frame* are synonymous. The laws of Special Relativity hold in inertial frames and therefore in the neighborhood of a *freely falling frame*. In this way the relativity principle is extended to arbitrary gravitational fields.

Associated with a given spacetime event there are an infinity of locally inertial frames related by Lorentz transformations. All are equivalent for the description of physical phenomena in a sufficiently small region of spacetime. So we arrive at a statement of the equivalence principle:

> At every spacetime point in an arbitrary gravitational field (meaning anytime and anywhere in the universe), a local inertial (Lorentz) frame can be chosen so that the laws of physics take on the form they have in Special Relativity.

This is the meaning of the equality of inertial and gravitational masses that Einstein sought. The restricted validity of inertial frames to small localities of any event suggested the very fruitful analogy with local flatness on a curved surface.

Einstein went further than the above statement of the equivalence principle. He spoke of the laws of nature rather than just the laws of physics. It seems entirely plausible that the extension is true, but we deal here only with physics.

The equivalence principle has great power. It is the instrument by which all the special relativistic laws of physics—valid in a gravity-free universe—can be generalized to a gravity-filled universe. We shall see how Einstein was able to give dynamic meaning to the spacetime continuum as an integral part of the physical world quite unlike the conception of an absolute spacetime in which the rest of physical processes take place. In Einstein's own words

> ... it is contrary to the mode of thinking in science to conceive of a thing (the spacetime continuum) which acts itself, but which cannot be acted upon.[16]

### 3.1.1  PHOTON IN A GRAVITATIONAL FIELD

The trajectory of photons is bent by a gravitational field. Imagine a freely falling elevator in a constant gravitational field. Its walls constitute an inertial frame as guaranteed by the equivalence principle. Therefore, a photon (as for a free particle) directed from one wall to the opposite along a path parallel to the floor will arrive at the other wall at the same height from which it started. But relative to the Earth, the elevator has fallen during the transit time. Therefore the photon has been deflected toward the earth and follows a curved path as observed from a frame fixed on the Earth.

Employing the conservation of energy and Newtonian physics, Einstein reasoned that the gravitational field acts on photons. Let a photon be emitted from $z_1$ vertically upward to $z_2$, and only for simplicity, let the field be uniform. A device located at $z_2$ converts its energy on arrival to a particle of mass $m$ with perfect efficiency. The particle drops to $z_1$ where its energy is now $m + mgh$, where $g$ is the acceleration due to the uniform field. A device at $z_1$ converts it into a photon of the same energy as possessed by the particle. The photon again is directed to $z_2$. If the original (and each succeeding photon) does not lose energy $(h\nu)gh$ in climbing the gravitational field equal to the energy gained by the particle in dropping in the field, we would have a device that creates energy. By the law of conservation of energy Einstein discovered the gravitational redshift, commonly designated by the factor $z$ and equal in this case to $gh$. The shift in energy of a photon by falling (in this case blue-shifted) in the earth's gravitational field has been directly confirmed in an experiment performed by Pound and Rebka [24].

In the above discussion the equivalence principle entered when the photon's inertial mass $(h\nu)$ was used also as its gravitational mass in computing the gravitational work. One can also see the role of the equivalence principle by considering a pulse of light emitted over a distance $h$ along the axis of a spaceship in uniform acceleration $g$ in outer space. The time taken for the light to reach the detector is $t = h$ (we use units $G = c = 1$). The difference in velocity of the detector acquired during the light travel time is $v = gt = gh$, the Doppler shift $z$ in the detected light. This experiment, carried out in the gravity-free environment of a spaceship whose rockets produce an acceleration $g$, must yield the same result for the energy shift of the photon in a uniform gravitational field $g$ according to the equivalence principle. The Pound–Rebka–Snyder experiments can therefore be regarded as an experimental proof of the equivalence principle [24, 25].

We may regard a radiating atom as a clock, with each wave crest regarded as a tick of the clock. Imagine two identical atoms situated one at some height above the other in the gravitational field of the earth. Since, by dropping in the gravitational field, the light is blue-shifted when compared to the radiation of an identical atom (clock) at the bottom, the clock at the top is seen to be running faster than the one at the bottom. Therefore, identical clocks, stationary with respect to the earth, run at different rates according to their different heights above the earth. Time flows at different rates in different gravitational fields.

An experiment carried out by J.C. Hafele and R.E. Keating tested a combination of time dilation in Special Relativity and the General Relativistic effect of differing gravitational field on the flow of time [26]. "During October 1971, four cesium atomic beam clocks were flown on regularly scheduled commercial jet flights around the world twice, once eastward and once westward, to test Einstein's theory of relativity with macroscopic clocks. From the actual flight paths of each trip, the theory predicted that the flying clocks, compared with reference clocks at the U.S. Naval Observatory, should have lost $40 \pm 23$ nanoseconds during the eastward trip and should have gained $275 \pm 21$ nanoseconds during the westward trip ... Relative to the atomic time scale of the U.S. Naval Observatory, the flying clocks lost $59 \pm 10$ nanoseconds during the eastward trip and gained $273 \pm 7$ nanosecond during the westward trip, where the errors are the corresponding standard deviations. These results provide an "unambiguous empirical resolution of the famous clock 'paradox' with macroscopic clocks." The different time-lapse of the eastward and westward journeys of the clocks has to to with their motion relative to the Earth bound reference clock.

The above experiment proves that the so-called "twin paradox" is not a paradox at all. The twins will have aged differently if they have had different trips in differing gravitational fields, one on the ground and one traveling in an airplane at a height above ground. Time dilation of Special Relativity is of different nature, having to do with differing speeds of the clocks, and is established in high-energy particle accelerators and routinely employed in designing experiments. We shall encounter both phenomena in later chapters.

## 3.1.2 TIDAL GRAVITY

Einstein predicted that a clock near a massive body would run more slowly than an identical distant clock. In doing so he arrived at a hint of the deep connection of the structure of spacetime and gravity. Two parallel straight lines never meet in the gravity-free, flat spacetime of Minkowski. A single inertial frame would suffice to describe all of spacetime. In formulating the equivalence principle (knowing that gravitational fields are not uniform and constant but depend on the motion of gravitating bodies and the position where gravitational effects are experienced), Einstein understood that only in a suitably small locality of spacetime do the laws of Special Relativity hold. Gravitational effects will be observed on a larger scale. *Tidal gravity* refers to the deviation from uniformity of the gravitational field at nearby points.

These considerations led Einstein to the notion of spacetime curvature. Whatever the motion of a free body in an arbitrary gravitational field, it will follow a straight-line trajectory over any small locality as guaranteed by the equivalence principle. And in a gravity-endowed universe, free particles whose trajectories are parallel in a local inertial frame, will not remain parallel over a large region of spacetime. This has a striking analogy with the surface of a sphere on which two straight lines that are parallel over a small region *do* meet and cross. What if in fact the particles are freely falling in curved spacetime? In this way of thinking, the law that free particles move in straight lines remains true in an arbitrary gravitational field, thus obeying the principle of relativity in a larger sense. Any sufficiently small region of curved spacetime is locally flat. Such paths in curved spacetime that have the property of being locally straight are called geodesics.

### 3.1.3  CURVATURE OF SPACETIME

Let us now consider a thought experiment. Two nearby bodies released from rest above the earth follow parallel trajectories over a small region of their trajectories, as we know from the equivalence principle. But if holes were drilled in the earth through which the bodies could fall, the bodies would meet and cross at the earth's center. So there is clearly no single Minkowski spacetime that covers a large region or the whole region containing a massive body.

Einstein's view was that spacetime curvature caused the bodies to cross, bodies that in this curved spacetime were following straight line paths in every small locality, just as they would have done in the whole of Minkowski (flat) spacetime in the absence of gravitating bodies. The presence of gravitating bodies denies the existence of a global inertial frame. Spacetime can be flat everywhere only if there exists such a global frame. Hence, spacetime is curved by massive bodies. In their presence a test particle follows a geodesic path, one that is always locally straight. The concept of a "gravitational force" has been replaced by the curvature of spacetime, and the natural free motions of particles in it are defined by geodesics.

### 3.1.4  ENERGY CONSERVATION AND CURVATURE

Interestingly, the conservation of energy can also be used to inform us that spacetime is curved. Consider a static gravitational field. Let us conjecture that spacetime is flat so that the Minkowski metric holds; we will arrive at a contradiction.

Imagine the following experiment performed by observers and their apparatus at rest with respect to the gravitational field and their chosen Lorentz frame in the supposed flat spacetime of Minkowski. At a height $z_1$ in the field, let a monochromatic light signal be emitted upward a height $h$ to $z_2 = z_1 + h$. Let the pulse be emitted for a specific time $dt_1$ during which N wavelengths (or photons) are emitted. Let the time during which they are received at $z_2$ be measured as $dt_2$. (Because the spacetime is assumed to be described by the Minkowski metric and the source and receiver are at rest in the chosen frame, the proper times and coordinate times are equal.)

Because the field in the above experiment is static, the path in the z-t plane will have the same shape for both the beginning and ending of the pulse (as for each photon) as they trace their path in the Minkowski spacetime we postulate to hold. The trajectories will not be lines at 45 degrees because of the field, but the curved paths will be congruent; a translation in time will make the paths lie one upon the other. Therefore $d\tau_2 = dt_2 = dt_1 = d\tau_1$ will be measured at the stationary detector if spacetime is Minkowskian. In this case, the frequency (and hence the energy received at $z_2$) is the same as that sent from $z_1$. But this cannot be. The photons comprising the signal must lose energy in climbing the gravitational field (see Section 3.1.1).

The conjecture that spacetime in the presence of a gravitational field is Minkowskian must therefore be false. We conclude that the presence of the gravitational field has caused spacetime to be curved. Such a line of reasoning was first conceived by Schild [27, 28, 29].

## 3.2    Gravity

"I was sitting in a chair at the patent office at Bern when all of a sudden a thought occurred to me: 'If a person falls freely he will not feel his own weight.' I was startled. This simple thought had a deep impression on me. It impelled me toward a theory of gravitation." *A. Einstein* [30]

### 3.2.1    EINSTEIN'S DISCOVERY

Einstein's discovery of General Relativity is generally regarded as the greatest intellectual achievement of any one man. I'm not sure how many people who subscribe to this view actually understand the theory and all of the mathematics that was needed to construct it. In this chapter we set it down for all who wish to see and follow it. *Subtle is the Lord* by Abraham Pais is a marvelous book that discusses the history of Einstein's struggles on the path to his discovery [31].

### 3.2.2    PARTICLE MOTION IN AN ARBITRARY GRAVITATIONAL FIELD

As we have seen, massive bodies generate curvature. Galaxies, stars, and other bodies are in motion; therefore the curvature of spacetime is everywhere changing. For this reason there is no "prior geometry". There are no immutable reference frames to which events in spacetime can be referred. Indeed, the changing geometry of spacetime and of the motion and arrangement of mass-energy in spacetime are inseparable parts of the description of physical processes. This is a very different idea of space and time from that of Newton and even of the Special Theory of Relativity. We now take up the unified discussion of gravitating matter and motion.

The power of the equivalence principle in informing us so simply that spacetime must be curved by the presence of massive bodies in the universe suggests a fruitful way of beginning. Following Weinberg [32], or indeed, following the notion expressed by Einstein in the quotation above, we seek the connection between an arbitrary reference frame and a reference frame that is freely falling with a particle that is moving only under the influence of an arbitrary gravitational field. In this freely falling and therefore locally inertial frame, the particle moves in a straight line. Denote the coordinates by $\xi^\alpha$. The equations of motion are

$$\frac{d^2\xi^\alpha}{d\tau^2} = 0 \, , \tag{3.35}$$

and the invariant interval (or proper time) between two neighboring spacetime events expressed in this frame, from (2.8), is

$$d\tau^2 = \eta_{\alpha\beta}d\xi^\alpha d\xi^\beta. \tag{3.36}$$

The freely falling coordinates may be regarded as functions of the co-ordinates $x^\mu$ of any arbitrary reference frame—curvilinear, accelerated, or rotating. We seek the connection between the equations of motion in the freely falling frame and the arbitrary one which, for example, might be the laboratory frame. From the chain rule for differentiation we can rewrite (3.35) as

$$0 = \frac{d}{d\tau}\left(\frac{\partial \xi^\alpha}{\partial x^\mu}\frac{dx^\mu}{d\tau}\right)$$
$$= \frac{\partial \xi^\alpha}{\partial x^\mu}\frac{d^2 x^\mu}{d\tau^2} + \frac{\partial^2 \xi^\alpha}{\partial x^\mu \partial x^\nu}\frac{dx^\mu}{d\tau}\frac{dx^\nu}{d\tau}.$$

Multiply by $\partial x^\lambda / \partial \xi^\alpha$, and use the chain rule again to obtain

$$\frac{dx^\lambda}{dx^\mu} = \frac{\partial x^\lambda}{\partial \xi^\alpha}\frac{\partial \xi^\alpha}{\partial x^\mu} = \delta^\lambda_\mu. \tag{3.37}$$

The equation of motion of the particle in an arbitrary frame when the particle is moving in an arbitrary gravitational field therefore is

$$\frac{d^2 x^\lambda}{d\tau^2} + \Gamma^\lambda_{\mu\nu}\frac{dx^\mu}{d\tau}\frac{dx^\nu}{d\tau} = 0. \tag{3.38}$$

Here $\Gamma^\lambda_{\mu\nu}$, defined by

$$\Gamma^\lambda_{\mu\nu} \equiv \frac{\partial x^\lambda}{\partial \xi^\alpha}\frac{\partial^2 \xi^\alpha}{\partial x^\mu \partial x^\nu}, \tag{3.39}$$

is called the *affine connection*. The affine connection is symmetric in its lower indices.

The path defined by equation (3.38) is called a *geodesic*, the extremal path in the spacetime of an arbitrary gravitational field. We do not see here that it is an extremal, but this is hinted at inasmuch as it defines the same path of (3.35), the straight-line path of a free particle as observed from its freely falling frame. In the next section we will see that a geodesic path is locally a straight line.

The invariant interval (3.36) can also be expressed in the arbitrary frame by writing $d\xi^\alpha = (\partial \xi^\alpha / \partial x^\mu)dx^\mu$ so that

$$d\tau^2 = g_{\mu\nu}dx^\mu dx^\nu \tag{3.40}$$

with

$$g_{\mu\nu} = \frac{\partial \xi^\alpha}{\partial x^\mu}\frac{\partial \xi^\beta}{\partial x^\nu}\eta_{\alpha\beta}. \tag{3.41}$$

In the new and arbitrary reference frame, the second term of (3.38) causes a deviation from a straight-line motion of the particle in this frame. There-fore, the second term represents the effect of the gravitational field. (To be sure, the connection coefficients also represent any other noninertial effects that may have been introduced by the choice of reference frame, such as rotation.)

The affine connection (3.39) appearing in the geodesic equation clearly plays an important role in gravity, and we study it further. We first show that the affine connection is a nontensor, and then show how it can be expressed in terms of the metric tensor and its derivatives. In this sense the metric behaves as the gravitational potential and the affine connection as the force. Write $\Gamma^{\lambda}_{\mu\nu}$ expressed in (3.39) in another coordinate system $x'^{\mu}$ and use the chain rule several times to rewrite it:

$$
\begin{aligned}
\Gamma'^{\lambda}_{\mu\nu} &= \frac{\partial x'^{\lambda}}{\partial x^{\rho}} \frac{\partial x^{\rho}}{\partial \xi^{\alpha}} \frac{\partial}{\partial x'^{\mu}} \left( \frac{\partial x^{\sigma}}{\partial x'^{\nu}} \frac{\partial \xi^{\alpha}}{\partial x^{\sigma}} \right) \\
&= \frac{\partial x'^{\lambda}}{\partial x^{\rho}} \frac{\partial x^{\rho}}{\partial \xi^{\alpha}} \left[ \frac{\partial x^{\sigma}}{\partial x'^{\nu}} \frac{\partial x^{\tau}}{\partial x'^{\mu}} \frac{\partial^2 \xi^{\alpha}}{\partial x^{\tau} \partial x^{\sigma}} + \frac{\partial^2 x^{\sigma}}{\partial x'^{\mu} \partial x'^{\nu}} \frac{\partial \xi^{\alpha}}{\partial x^{\sigma}} \right] \\
&= \frac{\partial x'^{\lambda}}{\partial x^{\rho}} \frac{\partial x^{\tau}}{\partial x'^{\mu}} \frac{\partial x^{\sigma}}{\partial x'^{\nu}} \Gamma^{\rho}_{\tau\sigma} + \frac{\partial x'^{\lambda}}{\partial x^{\rho}} \frac{\partial^2 x^{\rho}}{\partial x'^{\mu} \partial x'^{\nu}} \ . \quad (3.42)
\end{aligned}
$$

According to the transformation laws of tensors developed in Section 3.1, the second term on the right spoils the transformation law of the affine connection. It is therefore a nontensor.

Let us now obtain the expression of the affine connection in terms of the derivatives of the metric tensor. Form the derivative of (3.31):

$$
\frac{\partial}{\partial x'^{\kappa}} g'_{\mu\nu} = \frac{\partial}{\partial x'^{\kappa}} \left( g_{\rho\sigma} \frac{\partial x^{\rho}}{\partial x'^{\mu}} \frac{\partial x^{\sigma}}{\partial x'^{\nu}} \right) \ .
$$

Take the derivatives and form the following combination and find that it is equal to the above derivative:

$$
\begin{aligned}
\frac{\partial g'_{\kappa\nu}}{\partial x'^{\mu}} + \frac{\partial g'_{\kappa\mu}}{\partial x'^{\nu}} - \frac{\partial g'_{\mu\nu}}{\partial x'^{\kappa}} &= \frac{\partial x^{\tau}}{\partial x'^{\kappa}} \frac{\partial x^{\rho}}{\partial x'^{\mu}} \frac{\partial x^{\sigma}}{\partial x'^{\nu}} \left( \frac{\partial g_{\sigma\tau}}{\partial x^{\rho}} + \frac{\partial g_{\rho\tau}}{\partial x^{\sigma}} - \frac{\partial g_{\rho\sigma}}{\partial x^{\tau}} \right) \\
&\quad + 2 g_{\rho\sigma} \frac{\partial x^{\sigma}}{\partial x'^{\kappa}} \frac{\partial^2 x^{\rho}}{\partial x'^{\mu} \partial x'^{\nu}} \ .
\end{aligned}
$$

Multiply this equation by $\frac{1}{2}$ and then multiply the left and right sides by the left and right sides, respectively, of the law of transformation (3.32), namely,

$$
g'^{\lambda\kappa} = g^{\alpha\beta} \frac{\partial x'^{\lambda}}{\partial x^{\alpha}} \frac{\partial x'^{\kappa}}{\partial x^{\beta}} \ .
$$

Use the chain rule and rename several dummy indices to obtain

$$
\left\{ \begin{matrix} \lambda \\ \mu\nu \end{matrix} \right\}' = \frac{\partial x'^{\lambda}}{\partial x^{\rho}} \frac{\partial x^{\tau}}{\partial x'^{\mu}} \frac{\partial x^{\sigma}}{\partial x'^{\nu}} \left\{ \begin{matrix} \rho \\ \tau\sigma \end{matrix} \right\} + \frac{\partial x'^{\lambda}}{\partial x^{\rho}} \frac{\partial^2 x^{\rho}}{\partial x'^{\mu} \partial x'^{\nu}} \ , \quad (3.43)
$$

where the prime on {} means that it is evaluated in the $x'^\mu$ frame and the symbol stands for

$$\left\{ \begin{matrix} \lambda \\ \mu\nu \end{matrix} \right\} = \tfrac{1}{2} g^{\lambda\kappa} \left[ \frac{\partial g_{\kappa\nu}}{\partial x^\mu} + \frac{\partial g_{\kappa\mu}}{\partial x^\nu} - \frac{\partial g_{\mu\nu}}{\partial x^\kappa} \right]. \tag{3.44}$$

This is called a *Christoffel symbol of the second kind*. It is seen to transform in exactly the same way as the affine connection (3.42). Subtract the two to obtain

$$\left[ \Gamma^\lambda_{\mu\nu} - \left\{ \begin{matrix} \lambda \\ \mu\nu \end{matrix} \right\} \right]' = \frac{\partial x'^\lambda}{\partial x^\rho} \frac{\partial x^\tau}{\partial x'^\mu} \frac{\partial x^\sigma}{\partial x'^\nu} \left[ \Gamma^\rho_{\tau\sigma} - \left\{ \begin{matrix} \rho \\ \tau\sigma \end{matrix} \right\} \right]. \tag{3.45}$$

This shows that the difference is a tensor. According to the equivalence principle, at anyplace and anytime there is a local inertial frame $\xi^\alpha$ in which the effects of gravitation are absent, the metric is given by (2.9), and $\Gamma^\lambda_{\mu\nu}$ vanishes (compare (3.35) and (3.38)). Because the first derivatives of the metric tensor vanish in such a local inertial system, the Christoffel symbol also vanishes. Because the difference of the affine connection and the Christoffel symbol is a tensor which vanishes in this frame, the difference vanishes in all reference frames. So everywhere we find

$$\Gamma^\lambda_{\mu\nu} = \left\{ \begin{matrix} \lambda \\ \mu\nu \end{matrix} \right\} = \tfrac{1}{2} g^{\lambda\kappa} \left( g_{\kappa\nu,\mu} + g_{\kappa\mu,\nu} - g_{\mu\nu,\kappa} \right). \tag{3.46}$$

We use the "comma subscript" notation introduced earlier to denote differentiation (3.24).

Sometimes it is useful to have the superscript lowered on the affine connection

$$\Gamma_{\kappa\mu\nu} = g_{\kappa\lambda} \Gamma^\lambda_{\mu\nu}. \tag{3.47}$$

It is equal to the Christoffel symbol of the first kind

$$\Gamma_{\kappa\mu\nu} = \left[ \begin{matrix} \kappa \\ \mu\nu \end{matrix} \right] = \tfrac{1}{2} \left( g_{\kappa\nu,\mu} + g_{\kappa\mu,\nu} - g_{\mu\nu,\kappa} \right). \tag{3.48}$$

The above formulas provide a means of computing the affine connection from the derivatives of the metric tensor and will prove very useful. It is trivial from the above to prove that

$$\Gamma_{\kappa\mu\nu} + \Gamma_{\mu\kappa\nu} = g_{\mu\kappa,\nu}. \tag{3.49}$$

### 3.2.3 MATHEMATICAL DEFINITION OF LOCAL LORENTZ FRAMES

Spacetime is curved globally by the massive bodies in the universe. Therefore, we need to define mathematically the meaning of    "local Lorentz frame". In a rectilinear Lorentz frame the metric tensor is $\eta_{\mu\nu}$ (2.9). Therefore, in the local region around an event $P$ (a point in the four-dimensional spacetime continuum), the metric tensor, its coordinate derivatives, and the affine connection have the following values:

$$g_{\mu\nu}(P) = \eta_{\mu\nu}, \quad g_{\mu\nu,\alpha}(P) = 0, \quad \Gamma^\lambda_{\mu\nu}(P) = 0. \tag{3.50}$$

The third of these equations follows from the second and from (3.46). All local effects of gravitation disappear in such a frame. The geodesic equation (3.38) defining the path followed by a free particle in an arbitrary gravitational field becomes locally the equation of a uniform straight line, in accord with the equivalence principle.

Of course, physical measurements are always subject to the precision of the measuring devices. The extent of the local region around $P$, in which the above equations will hold and in which spacetime is said to be flat, will depend on the accuracy of the devices and therefore their ability to detect deviations from the above conditions as one measures further from $P$.

### 3.2.4   GEODESICS

In the Special Theory of Relativity a free particle remains at rest or moves with constant velocity in a straight line. A straight line is the shortest distance between two points in Euclidean three-dimensional space. In Minkowski spacetime a straight line is the longest interval between two events, as we shall shortly see. Both situations are covered by saying that a straight line is an extremal path between two points. We shall show that in an arbitrary gravitational field, a particle moving under the influence gravity alone, follows a path that is, in the sense that we shall define, the straightest line possible in curved spacetime.

We first show that a straight-line path between two events in Minkowski spacetime maximizes the proper time. This is easily proved. Orient the axis so that the two events marking the ends of the path, $A$ and $B$, lie on the t-axis with coordinates $(0,0,0,0)$ and $(T,0,0,0)$, and consider an alternate path in the t-x plane that consists of two straight-line segments that pass from $A$ to $B$ through $(T/2, R/2, 0, 0)$. The proper time as measured on the second path is

$$\tau = 2\sqrt{(T/2)^2 - (R/2)^2} = \sqrt{T^2 - R^2}. \tag{3.51}$$

For any finite $R$, $\tau$ is smaller than the proper time along the straight-line path from $A$ to $B$, namely, $T$. Therefore, a straight-line path is a maximum in proper time.

We have referred to the equation of motion of a particle moving freely in an arbitrary gravitational field (3.38) as a geodesic equation. The geodesic of a photon is null; $d\tau = 0$. In general, a geodesic that is not null is the extremal path of

$$\int_A^B d\tau \qquad (3.52)$$

where $A$ and $B$ refer to spacetime events on the geodesic. To prove this result, let $x^\mu(\tau)$ denote the coordinates along the geodesic path, parameterized by the proper time, and let $x^\mu(\tau) + \delta x^\mu(\tau)$ denote a neighboring path with the same end points, $A$ to $B$. From

$$d\tau^2 = g_{\mu\nu} dx^\mu dx^\nu , \qquad (3.53)$$

we have to first order in the variation,

$$2d\tau\,\delta(d\tau) = \delta g_{\mu\nu} dx^\mu dx^\nu + 2g_{\mu\nu} dx^\mu \delta(dx^\nu)$$
$$= dx^\mu dx^\nu g_{\mu\nu,\lambda} \delta x^\lambda + 2g_{\mu\nu} dx^\mu d(\delta x^\nu) . \qquad (3.54)$$

Recalling the four-velocity, $u^\mu = dx^\mu/d\tau$, we have

$$\delta(d\tau) = \left( \tfrac{1}{2} u^\mu u^\nu g_{\mu\nu,\lambda} \delta x^\lambda + g_{\mu\lambda} u^\mu \frac{d}{d\tau} \delta x^\lambda \right) d\tau . \qquad (3.55)$$

Thus

$$\delta \int_A^B d\tau = \int_A^B \left[ \tfrac{1}{2} u^\mu u^\nu g_{\mu\nu,\lambda} - \frac{d}{d\tau}\left( g_{\mu\lambda} u^\mu \right) \right] \delta x^\lambda \, d\tau \qquad (3.56)$$

where an integration by parts in the second term was performed. Because the variation of the path $\delta x^\lambda$ is arbitrary save for its end points being zero, we obtain as the extremal condition,

$$\frac{d}{d\tau}\left( g_{\mu\lambda} u^\mu \right) - \tfrac{1}{2} u^\mu u^\nu g_{\mu\nu,\lambda} = 0 . \qquad (3.57)$$

The first and second terms can be rewritten:

$$\frac{d}{d\tau}\left( g_{\mu\lambda} u^\mu \right) = g_{\mu\lambda} \frac{du^\mu}{d\tau} + g_{\mu\lambda,\nu} u^\mu u^\nu ,$$
$$g_{\mu\lambda,\nu} u^\mu u^\nu = \tfrac{1}{2}(g_{\mu\lambda,\nu} + g_{\lambda\nu,\mu}) u^\mu u^\nu . \qquad (3.58)$$

Now using the relationship (3.48), we find

$$g_{\mu\lambda} \frac{du^\mu}{d\tau} + \Gamma_{\lambda\mu\nu} u^\mu u^\nu = 0 . \qquad (3.59)$$

Multiplying by $g^{\sigma\lambda}$ and summing on $\lambda$, we obtain the geodesic equation (3.38):

$$\frac{du^\sigma}{d\tau} + \Gamma^\sigma_{\mu\nu} u^\mu u^\nu = 0 . \qquad (3.60)$$

This completes the proof that the path defined by the geodesic equation, the equation of motion of a particle in a purely gravitational field, extremizes the proper time between any two events on the path.

The straight-line path between two events in Minkowski spacetime maximizes the interval between the events. We proved that a geodesic path, in the general case that a gravitational field is present, will be an extremum, but if the spacetime separation of the ends of the path is large, there may be two geodesic paths, one of minimum and one of maximum length. The geodesic path of a particle in spacetime is frequently referred to as its *world line*. A world line is a continuous sequence of points in spacetime; it represent the history of a particle or photon, or in general, an *event*.

In a region of spacetime sufficiently small that the Minkowski metric holds (the existence of which locality is guaranteed by the equivalence principle), we see that the geodesic equation reduces to that for uniform straight-line motion,

$$\frac{du^\mu}{d\tau} = 0 \,. \tag{3.61}$$

Therefore, the path of a particle moving under the influence of a general gravitational field will be locally straight. But we know that no global Lorentz frame exists in the presence of gravitating bodies; therefore, geodesic paths will in general be curved. However, in the above sense they will be as straight as possible in curved spacetime.

## 3.2.5   COMPARISON WITH NEWTON'S GRAVITY

We confirm the assertion made earlier that the metric tensor $g_{\mu\nu}$ takes the place in General Relativity that the Newtonian potential occupies in Newton's theory. Of course this must be done in a weak field situation for it is only there that Newton's theory applies. For this reason, of the ten independent $g_{\mu\nu}$'s, only one can be involved in the correspondence.

We consider a particle moving slowly in a weak static gravitational field. From the Special Theory of Relativity we have

$$d\tau = \left(dt^2 - d\mathbf{r}^2\right)^{1/2} = \left(1 - \mathbf{v}^2\right)^{1/2}dt, \qquad \mathbf{v} = \frac{d\mathbf{r}}{dt}, \tag{3.62}$$

where boldface symbols denote three-vectors. The slowly moving assumption is

$$\left|\frac{d\mathbf{r}}{dt}\right| << \frac{dt}{d\tau} \approx 1 \,. \tag{3.63}$$

(Recall that we use units such that $c = 1$.) So the geodesic equation (3.38) can be written with the neglect of the velocity terms as

$$\frac{d^2x^\mu}{d\tau^2} + \Gamma^\mu_{00}\left(\frac{dt}{d\tau}\right)^2 = 0 \,. \tag{3.64}$$

Because the field is static, the time derivatives of $g_{\mu\nu}$ vanish. Consequently,

$$\Gamma^{\mu}_{00} = \tfrac{1}{2}g^{\mu\nu}\left(2g_{\nu 0,0} - g_{00,\nu}\right) = -\tfrac{1}{2}g^{\mu\nu}g_{00,\nu}(1 - \delta^{\nu}_0)\,. \tag{3.65}$$

Because the field is weak we may take

$$g_{00} = (1 + \delta)\eta_{00}\,, \tag{3.66}$$

where $\delta \ll 1$. To first order in the small quantities, we have

$$\Gamma^{\mu}_{00} = -\tfrac{1}{2}\eta^{\mu\nu}\eta_{00}\frac{d\delta}{dx^{\nu}}(1 - \delta^{\nu}_0)\,. \tag{3.67}$$

Thus the geodesic equations become

$$\frac{d^2\mathbf{r}}{d\tau^2} = -\tfrac{1}{2}\left(\frac{dt}{d\tau}\right)^2\nabla\delta, \qquad \frac{d^2t}{d\tau^2} = 0\,. \tag{3.68}$$

The second of these tells us that $\tau = at + b$. So we may write the first as

$$\frac{d^2\mathbf{r}}{dt^2} = -\tfrac{1}{2}\nabla\delta\,. \tag{3.69}$$

Newton's equation is

$$\frac{d^2\mathbf{r}}{dt^2} = -\nabla V\,, \tag{3.70}$$

where $V$ is the gravitational potential. Comparing, we have

$$g_{00} = 1 + 2V\,. \tag{3.71}$$

In particular, if the gravitational field is produced by a body of mass $M$,

$$V = -\frac{GM}{r} \quad\Longrightarrow\quad g_{00} = 1 - \frac{2GM}{r}\,, \tag{3.72}$$

where $G$ is Newton's constant. Thus we see for weak fields how the metric is related to the Newtonian potential.

## 3.3   Covariance

### 3.3.1   Principle of general covariance

Physical laws in their form ought to be independent of the frame in which they are expressed and of the location in the universe, that is, independent of the gravitational field. The principle of general covariance states that a law of physics holds in a general gravitational field if it holds in the absence of gravity and its form is invariant to any coordinate transformation.

Physical laws frequently involve space-time derivatives of scalars, vectors, or tensors. We have seen that the derivative of a scalar is a vector but that the ordinary derivative of a vector or a tensor is not a tensor (page 25). Therefore, we need a type of derivative—a covariant derivative—that reduces to ordinary differentiation in the absence of gravity and which retains its form under any coordinate transformation, that is, in any gravitational field.

### 3.3.2  COVARIANT DIFFERENTIATION

Take the derivative of the expression of the covariant vector transformation law (3.14),

$$\frac{dA'_\mu}{dx'^\rho} = \frac{\partial x^\nu}{\partial x'^\mu}\frac{\partial x^\sigma}{\partial x'^\rho}\frac{\partial A_\nu}{\partial x^\sigma} + \frac{\partial^2 x^\nu}{\partial x'^\rho \partial x'^\mu}A_\nu \ .$$

If only the first term were present we would have the correct transformation law for a covariant tensor. Now multiply the left and right sides of (3.42) by the left and right sides of (3.14), respectively, and rearrange to find

$$\Gamma'^\lambda_{\mu\nu} A'_\lambda = \frac{\partial x^\alpha}{\partial x'^\mu}\frac{\partial x^\beta}{\partial x'^\nu}\Gamma^\kappa_{\alpha\beta}A_\kappa + \frac{\partial^2 x^\kappa}{\partial x'^\mu \partial x'^\nu}A_\kappa \ .$$

Subtracting the above two equations after renaming dummy indices of summation, we get

$$\left(\frac{dA'_\mu}{dx'^\nu} - \Gamma'^\lambda_{\mu\nu}A'_\lambda\right) = \frac{\partial x^\alpha}{\partial x'^\mu}\frac{\partial x^\beta}{\partial x'^\nu}\left(\frac{dA_\alpha}{dx^\beta} - \Gamma^\lambda_{\alpha\beta}A_\lambda\right) , \tag{3.73}$$

which proves the tensor character of the quantity in brackets. This we call the *covariant derivative* of a covariant vector. We denote it by

$$A_{\mu;\nu} \equiv \frac{dA_\mu}{dx^\nu} - \Gamma^\lambda_{\mu\nu}A_\lambda , \tag{3.74}$$

and the "semicolon subscript" shall denote the covariant derivative, and imply the operations shown on the right. The covariant derivative of a covariant vector is a second-rank covariant tensor which reduces to ordinary differentiation in inertial frames—and therefore locally in any gravitational field.

Through similar manipulations we find the covariant derivative of a contravariant vector,

$$A^\mu_{\ ;\nu} \equiv \frac{dA^\mu}{dx^\nu} + \Gamma^\mu_{\sigma\nu}A^\sigma \ . \tag{3.75}$$

This is a second-rank mixed tensor because its transformation law is

$$\left(\frac{dA'^\mu}{dx'^\nu} + \Gamma'^\mu_{\lambda\nu}A'^\lambda\right) = \frac{\partial x'^\mu}{\partial x^\alpha}\frac{\partial x^\beta}{\partial x'^\nu}\left(\frac{dA^\alpha}{dx^\beta} + \Gamma^\alpha_{\kappa\beta}A^\kappa\right) . \tag{3.76}$$

The covariant derivative of a mixed tensor of arbitrary order can be obtained by successive application of the above two rules to each index; there is one ordinary derivative of the tensor and an affine connection for each index with sign as indicated by the above.

In particular, the covariant derivative of the metric tensor is

$$g_{\mu\nu;\lambda} = \frac{\partial g_{\mu\nu}}{\partial x^\lambda} - \Gamma^\rho_{\mu\lambda} g_{\rho\nu} - \Gamma^\rho_{\nu\lambda} g_{\rho\mu}\,. \tag{3.77}$$

In a local inertial frame, where the affine connection and the derivative of the metric tensor vanish, we see that the covariant derivative of the metric tensor vanishes in that frame. But because this itself is a tensor, it must vanish in all frames. Similarly, for the covariant derivative of $g^{\mu\nu}$,

$$g_{\mu\nu;\lambda} = 0 = g^{\mu\nu}{}_{;\lambda}\,. \tag{3.78}$$

### 3.3.3  GEODESIC EQUATION FROM COVARIANCE PRINCIPLE

As an important example of the application of the covariant derivative, consider the four-velocity of a free particle in a Lorentz frame in the absence of gravity. We denote the four-velocity by $u^\mu = dx^\mu/d\tau$ and its equation of motion is $du^\mu/d\tau = 0$, or equivalently in differential form,

$$du^\mu = 0\,. \tag{3.79}$$

The covariant derivative (3.76) was introduced to preserve the vector or tensor character so that a law expressed in such form is preserved in form for all coordinate transformations in accord with the principle of relativity. The equation expressing the law is said to be covariant if its form is preserved. Therefore the law of free motion (3.79) in a Lorentz frame in the absence of gravity is generalized to frames in arbitrary gravitational fields by requiring that the covariant differential of the four-velocity vanish:

$$\begin{aligned}
0 = u^\mu{}_{;\nu} dx^\nu &= \frac{du^\mu}{dx^\nu} dx^\nu + \Gamma^\mu_{\sigma\nu} u^\sigma dx^\nu \\
&= du^\mu + \Gamma^\mu_{\sigma\nu} u^\sigma dx^\nu\,.
\end{aligned} \tag{3.80}$$

Dividing the above equation by $(d\tau)^2$ yields the expected result—the geodesic equation (3.38)—the equation of motion derived previously for a free particle in an arbitrary gravitational field:

$$\frac{d^2 x^\lambda}{d\tau^2} + \Gamma^\lambda_{\mu\nu} \frac{dx^\mu}{d\tau} \frac{dx^\nu}{d\tau} = 0\,. \tag{3.81}$$

This is an example of the application of the principle of general covariance and it is seen to rest on the equivalence principle, which assures us that a Lorentz frame can be erected locally.

To restate the principle briefly, *any law that holds in the special theory of relativity and in the absence of gravity can be generalized by replacing the metric $\eta_{\mu\nu}$ by $g_{\mu\nu}$ and replacing ordinary derivatives by covariant derivatives.*

We obtain an additional result that we need later, namely, the equations of motion for the covariant components of the four-velocity. The law of motion of a free particle in the special theory, expressed in differential form as in (3.79), implies at once that $du_\mu = g_{\mu\nu} du^\nu = 0$. The covariant translation of this fact is

$$0 = u_{\mu;\nu} dx^\nu = \frac{du_\mu}{dx^\nu} dx^\nu - \Gamma^\lambda_{\mu\nu} u_\lambda dx^\nu \tag{3.82}$$

or

$$\frac{d^2 x_\mu}{d\tau^2} - \Gamma^\lambda_{\mu\nu} \frac{dx_\lambda}{d\tau} \frac{dx^\nu}{d\tau} = 0 . \tag{3.83}$$

This is the equation corresponding to (3.81) for the covariant acceleration. We carry the analysis a step further. Examine the second term on the left.

$$\begin{aligned}
\Gamma^\lambda_{\mu\nu} u_\lambda u^\nu &= \frac{1}{2} g^{\lambda\kappa} \left( g_{\kappa\nu,\mu} + g_{\kappa\mu,\nu} - g_{\mu\nu,\kappa} \right) u_\lambda u^\nu \\
&= \frac{1}{2} \left( g_{\kappa\nu,\mu} + g_{\kappa\mu,\nu} - g_{\mu\nu,\kappa} \right) u^\kappa u^\nu .
\end{aligned} \tag{3.84}$$

Because of the symmetry of the product $u^\kappa u^\nu$, the last two terms in the bracket cancel. We are left with

$$\frac{du_\mu}{d\tau} = \frac{1}{2} g_{\kappa\nu,\mu} u^\kappa u^\nu . \tag{3.85}$$

This proves that if all the $g_{\alpha\beta}$ are independent of some coordinate component, say $x^\mu$, then the covariant velocity $u_\mu$ is a constant along the particle's trajectory. This result can be used to demonstrate that in the vicinity of a rotating body, local inertial frames are dragged in the direction of rotation so that a body dropped freely from a great distance falls, not toward the bodies center, but is dragged ever more strongly in the sense of the rotation [4].

### 3.3.4  COVARIANT DIVERGENCE AND CONSERVED QUANTITIES

The element of four-volume transforms under coordinate change as

$$dx'^0 dx'^1 dx'^2 dx'^3 = J \, dx^0 dx^1 dx^2 dx^3 , \tag{3.86}$$

where $J$ is the Jacobian of the transformation,

$$J = \det \left| \frac{\partial x'^\rho}{\partial x^\mu} \right| . \tag{3.87}$$

For brevity the four-volume element is often written $d^4x$.

The transformation law for the metric tensor is

$$g_{\mu\nu} = \frac{\partial x'^\alpha}{\partial x^\mu} g'_{\alpha\beta} \frac{\partial x'^\beta}{\partial x^\nu} . \tag{3.88}$$

We may regard this as an element in the product of three matrices. The corresponding determinant equation is

$$g = Jg'J = J^2g' \tag{3.89}$$

where $g = \det|g_{\mu\nu}|$ and is a negative quantity as can be verified by looking at the Minkowski metric. Thus, we may write

$$\sqrt{-g} = J\sqrt{-g'} . \tag{3.90}$$

If $S = S'$ is a scalar field, then

$$\int_{V_4} S\sqrt{-g}\, d^4x = \int_{V_4} S\sqrt{-g'} J\, d^4x = \int_{V_4} S'\sqrt{-g'}\, d^4x' \tag{3.91}$$

is an invariant where $V_4$ is a prescribed four-volume. The quantity

$$\mathcal{S} \equiv S\sqrt{-g} \tag{3.92}$$

is called a *scalar density*, and its integral over a region of spacetime is invariant to a coordinate transformation. Also, and very important to us, $\sqrt{-g}\, d^4x$ is the invariant volume element.

The covariant derivative of a vector $A^\mu$ is given by (3.75). If we contract indices, according to (3.28) we have a scalar. This is the *covariant divergence* of $A^\mu$:

$$A^\mu_{\;;\mu} \equiv A^\mu_{\;,\mu} + \Gamma^\mu_{\nu\mu} A^\nu . \tag{3.93}$$

From (3.46) we find

$$\Gamma^\nu_{\mu\nu} = \tfrac{1}{2} g^{\nu\kappa} \left( g_{\kappa\nu,\mu} + g_{\kappa\mu,\nu} - g_{\mu\nu,\kappa} \right) . \tag{3.94}$$

Interchange the names of the dummy summation indices in the second term on the right to see that it cancels the third. Thus

$$\Gamma^\nu_{\mu\nu} = \tfrac{1}{2} g^{\nu\kappa} g_{\kappa\nu,\mu} . \tag{3.95}$$

We need still another result. Denote the cofactor of the element $g_{\alpha\beta}$ by $C^{\alpha\beta}$. The determinant $g = \det|g_{\alpha\beta}|$ can be expanded in any of the set of minors (i.e., any $\alpha = 0, 1, 2,$ or $3$) in the equation

$$g = g_{(\alpha)\beta} C^{(\alpha)\beta} \quad \text{(no sum on } \alpha\text{)}. \tag{3.96}$$

Because the cofactor contains no elements $g_{(\alpha)\beta}$, we find

$$\frac{\partial g}{\partial g_{\alpha\nu}} = \frac{\partial(g_{(\alpha)\beta}C^{(\alpha)\beta})}{\partial g_{\alpha\nu}} = \frac{\partial g_{\alpha\mu}}{\partial g_{\alpha\nu}}C^{\alpha\mu} = \delta^\nu_\mu C^{\alpha\mu} = C^{\alpha\nu}. \tag{3.97}$$

Therefore,

$$g_{,\alpha} = \frac{\partial g}{\partial g_{\mu\nu}}g_{\mu\nu,\alpha} = C^{\mu\nu}g_{\mu\nu,\alpha}. \tag{3.98}$$

We need the expression

$$C^{\mu\nu} = gg^{\mu\nu}, \tag{3.99}$$

which can be proved by multiplying by $g_{\mu\nu}$ and summing only over $\nu$,

$$g_{(\mu)\nu}C^{(\mu)\nu} = g_{(\mu)\nu}g^{(\mu)\nu}g = g. \tag{3.100}$$

This is the determinant expansion in minors (3.96). Thus, we have derived the result

$$g_{,\alpha} = gg^{\mu\nu}g_{\mu\nu,\alpha}. \tag{3.101}$$

Hence,

$$\Gamma^\nu_{\mu\nu} = \tfrac{1}{2}g^{-1}g_{,\mu} = \tfrac{1}{2}\left(\ln(-g)\right)_{,\mu} = \frac{1}{\sqrt{-g}}\left(\sqrt{-g}\right)_{,\mu}. \tag{3.102}$$

We can use this to rewrite the covariant divergence of $A^\mu$ as

$$A^\mu_{\;;\mu} = \frac{1}{\sqrt{-g}}\left(\sqrt{-g}A^\mu\right)_{,\mu} \tag{3.103}$$

With (3.102) in (3.93), we obtain the important result for the covariant divergence,

$$\sqrt{-g}A^\mu_{\;;\mu} = \left(\sqrt{-g}A^\mu\right)_{,\mu}. \tag{3.104}$$

The left side is a scalar density. From the invariance of the integral of a scalar density over a prescribed four-volume, we have the invariant

$$\int_{V_4}\sqrt{-g}A^\mu_{\;;\mu}\,d^4x = \int_{V_4}\left(\sqrt{-g}A^\mu\right)_{,\mu}d^4x. \tag{3.105}$$

The right side can be converted to a surface integral over a three-volume at a definite time $x^0$ by Gauss' theorem.

If the covariant divergence vanishes, we get a conservation law as follows:

$$A^\mu_{\;;\mu} = 0 \quad \Longrightarrow \quad \left(\sqrt{-g}A^\mu\right)_{,\mu} = 0. \tag{3.106}$$

As a result, we obtain

$$\left(\sqrt{-g}A^0\right)_{,0} = -\left(\sqrt{-g}A^m\right)_{,m} \quad \text{(summed over } m = 1 - 3\text{)}. \quad (3.107)$$

Integrate the above expression over a three-volume at definite time $x^0$ to find

$$\frac{\partial}{\partial x^0} \int_V \sqrt{-g}A^0 \, d^3x = -\int_V \left(\sqrt{-g}A^m\right)_{,m} d^3x \qquad (3.108)$$

$$= -\int_S \sqrt{-g}\mathbf{A} \cdot d\mathbf{S}. \qquad (3.109)$$

If there is no three-current $\sqrt{-g}\mathbf{A}$ crossing the surface, then the quantity of density $\sqrt{-g}A^0$ contained within $V$ is constant,

$$\int_V \sqrt{-g}A^0 \, d^3x = \text{ constant} \qquad (3.110)$$

This quantity is frequently referred to as the total charge of whatever $A^\mu$ represents.

We can apply precisely the same reasoning to the covariant divergence of an antisymmetric tensor:

$$\text{If} \quad A^{\mu\nu} = -A^{\nu\mu}, \quad \text{then} \quad \sqrt{-g}A^{\mu\nu}{}_{;\nu} = \left(\sqrt{-g}A^{\mu\nu}\right)_{,\nu}, \qquad (3.111)$$

where the quantity on the left is a *vector density* according to the previous section. Similarly we can derive conservation laws for the three-volume integral of the four densities $\sqrt{-g}A^{\mu 0}$ if the covariant divergence vanishes and there is no three-flux through the surface of the volume. However, if the tensor is not antisymmetric, the above theorem does not generally apply in curved spacetime to a tensor of more than one index.

## 3.4   Riemann Curvature Tensor

The order of ordinary differentiation in flat spacetime does not matter. The order of covariant differentiation does matter in curved spacetime. From an investigation of this fact we arrive at a measure of curvature.

### 3.4.1   SECOND COVARIANT DERIVATIVE OF SCALARS AND VECTORS

If we take the covariant derivative of a scalar twice and then invert the order, the answer is easily verified to be the same:

$$S_{;\mu;\nu} = S_{;\mu,\nu} - \Gamma^\alpha_{\mu\nu}S_{;\alpha} = S_{,\mu,\nu} - \Gamma^\alpha_{\mu\nu}S_{;\alpha}, \qquad (3.112)$$

where we use the fact in the second equality that the covariant derivative of a scalar is the ordinary derivative $S_{;\mu} = S_{,\mu}$. The above result is symmetrical in $\mu, \nu$.

However for vectors and tensors, a changed order of differentiation in general produces a different result. The operations involved, all defined above, are many but straightforward. The result for the vector $A_\sigma$ is

$$A_{\sigma;\mu;\nu} - A_{\sigma;\nu;\mu} = A_\rho R^\rho_{\sigma\mu\nu}, \tag{3.113}$$

where

$$R^\rho_{\sigma\mu\nu} \equiv \Gamma^\rho_{\sigma\nu,\mu} - \Gamma^\rho_{\sigma\mu,\nu} + \Gamma^\alpha_{\sigma\nu}\Gamma^\rho_{\alpha\mu} - \Gamma^\alpha_{\sigma\mu}\Gamma^\rho_{\alpha\nu} \tag{3.114}$$

is the *Riemann–Christoffel* curvature tensor. We know that it is a tensor because the left side of (3.113) is a tensor and $A_\nu$ is any vector. Riemann is the only tensor that can be constructed from the metric tensor and its first and second derivatives (cf. Ref. [32], p. 133).

## 3.4.2  SYMMETRIES OF THE RIEMANN TENSOR

Riemann has a number of symmetry properties that can be easily derived from the above expression:

$$\begin{aligned} R^\mu_{\nu\rho\sigma} &= -R^\mu_{\nu\sigma\rho}, \\ R^\alpha_{\nu\rho\sigma} + R^\alpha_{\sigma\nu\rho} + R^\alpha_{\rho\sigma\nu} &= 0. \end{aligned} \tag{3.115}$$

Lowering the index on the Riemann tensor, we get

$$R_{\rho\sigma\mu\nu} = g_{\rho\alpha}R^\alpha_{\sigma\mu\nu}. \tag{3.116}$$

The additional symmetries follow:

$$\begin{aligned} R_{\mu\nu\rho\sigma} &= -R_{\nu\mu\rho\sigma} = -R_{\mu\nu\sigma\rho}, \\ R_{\mu\nu\rho\sigma} &= R_{\rho\sigma\mu\nu} = R_{\sigma\rho\nu\mu}. \end{aligned} \tag{3.117}$$

As a consequence of the symmetries only 20 of the $4^4 = 256$ components of Riemann are independent. In two dimensions there are 15 such symmetry relationships. Consequently, there are $2^4 - 15 = 1$ independent components of the Riemann tensor, namely, the Gaussian curvature. (See Ref. [33] p. 60 and appendix B for a discussion of curvature in two dimensions.)

We shall encounter two additional objects that are obtained from the Riemann tensor, the *Ricci tensor*,

$$R_{\mu\nu} = R^\rho_{\mu\nu\rho}, \tag{3.118}$$

and the *scalar!curvature*,

$$R = g^{\mu\nu} R_{\mu\nu}. \tag{3.119}$$

Multiply the left and right side of (3.117) by $g^{\mu\sigma}$ and then rename indices to find

$$R_{\mu\nu} = R_{\nu\mu}. \tag{3.120}$$

Because of this symmetry, when we raise an index on the Ricci tensor, it is unnecessary to preserve the location,

$$R^{\mu}{}_{\nu} = R_{\nu}{}^{\mu} = R^{\mu}_{\nu}. \tag{3.121}$$

From the definition of the Ricci tensor in terms of the Riemann tensor, we have the following explicit expression:

$$R_{\mu\nu} = \Gamma^{\alpha}_{\mu\alpha,\nu} - \Gamma^{\alpha}_{\mu\nu,\alpha} - \Gamma^{\alpha}_{\mu\nu}\Gamma^{\beta}_{\alpha\beta} + \Gamma^{\alpha}_{\mu\beta}\Gamma^{\beta}_{\nu\alpha}. \tag{3.122}$$

The first term might appear to contradict the assertion that $R_{\mu\nu}$ is symmetric in $\mu, \nu$. However the result (3.102) proves that the Ricci tensor is symmetric.

### 3.4.3  TEST FOR FLATNESS

If spacetime is flat, then we may choose a rectilinear coordinate system in which case the metric tensor is a constant throughout spacetime. Then according to (3.46) the nontensor $\Gamma^{\lambda}_{\mu\nu}$ vanishes in this frame in all spacetime. So also do the derivatives of $\Gamma^{\lambda}_{\mu\nu}$. Therefore the Riemann tensor (3.114) vanishes everywhere at all times in flat spacetime. Because this is a statement about a tensor, it is true in any coordinate system, rectilinear or not. The converse is true but more difficult to prove: If Riemann vanishes, spacetime is flat. We prove this later in Section 3.6.3.

### 3.4.4  SECOND COVARIANT DERIVATIVE OF TENSORS

An arbitrary second-rank tensor can be expressed as the sum of products $A_{\mu}B_{\nu}$. It is simpler to start by examining the second covariant derivative of such a product:

$$
\begin{aligned}
(A_{\mu}B_{\nu})_{;\rho;\sigma} &= (A_{\mu;\rho}B_{\nu} + A_{\mu}B_{\nu;\rho})_{;\sigma} \\
&= A_{\mu;\rho;\sigma}B_{\nu} + A_{\mu}B_{\nu;\rho;\sigma} + A_{\mu;\rho}B_{\nu;\sigma} + A_{\mu;\sigma}B_{\nu;\rho}.
\end{aligned}
$$

Interchange $\rho, \sigma$, and subtract to find

$$
\begin{aligned}
(A_{\mu}B_{\nu})_{;\rho;\sigma} &= (A_{\mu}B_{\nu})_{;\sigma;\rho} \\
&= A_{\mu}(B_{\nu;\rho;\sigma} - B_{\nu;\sigma;\rho}) + (A_{\mu;\rho;\sigma} - A_{\mu;\sigma;\rho})B_{\nu} \\
&= A_{\mu}B_{\alpha}R^{\alpha}_{\nu\rho\sigma} + A_{\alpha}R^{\alpha}_{\mu\rho\sigma}B_{\nu}.
\end{aligned}
$$

We can form an arbitrary linear combination of such products of first-rank tensors to obtain the result for a general tensor,

$$T_{\mu\nu;\rho;\sigma} - T_{\mu\nu;\sigma;\rho} = T_{\mu\alpha}R^{\alpha}_{\nu\rho\sigma} + T_{\alpha\nu}R^{\alpha}_{\mu\rho\sigma}. \tag{3.123}$$

### 3.4.5  BIANCHI IDENTITIES

The Bianchi identities are extremely important for the further development of the theory of gravity, allowing us to prove that the Einstein tensor, which we come to next, has vanishing divergence.

Apply the above result to the particular case that the second-rank tensor is the covariant derivative of a vector $T_{\mu\nu} = A_{\mu;\nu}$,

$$A_{\mu;\nu;\rho;\sigma} - A_{\mu;\nu;\sigma;\rho} = A_{\mu;\alpha}R^{\alpha}_{\nu\rho\sigma} + A_{\alpha;\nu}R^{\alpha}_{\mu\rho\sigma} . \tag{3.124}$$

Now write down the additional two equations obtained from this by cyclic permutation of the indices $(\nu\rho\sigma)$, and add the three equations. First study the left side of the sum. Use (3.113) to get

$$\begin{aligned} \text{LHS} &= (A_{\mu;\nu;\rho} - A_{\mu;\rho;\nu})_{;\sigma} + (A_{\mu;\sigma;\nu} - A_{\mu;\nu;\sigma})_{;\rho} \\ +(A_{\mu;\rho;\sigma} - A_{\mu;\sigma;\rho})_{;\nu} &= (A_{\alpha}R^{\alpha}_{\mu\nu\rho})_{;\sigma} + (A_{\alpha}R^{\alpha}_{\mu\sigma\nu})_{;\rho} + (A_{\alpha}R^{\alpha}_{\mu\rho\sigma})_{;\nu} . \end{aligned}$$

Using (3.115) in the sum of the right-hand sides of the cyclic permutation, we are left with

$$\text{RHS} = A_{\alpha;\nu}R^{\alpha}_{\mu\rho\sigma} + A_{\alpha;\sigma}R^{\alpha}_{\mu\nu\rho} + A_{\alpha;\rho}R^{\alpha}_{\mu\sigma\nu} .$$

Equating left and right sides and canceling common terms, we find

$$A_{\alpha}(R^{\alpha}_{\mu\nu\rho;\sigma} + R^{\alpha}_{\mu\sigma\nu;\rho} + R^{\alpha}_{\mu\rho\sigma;\nu}) = 0 . \tag{3.125}$$

Because $A_{\alpha}$ is any vector,

$$R^{\alpha}_{\mu\nu\rho;\sigma} + R^{\alpha}_{\mu\sigma\nu;\rho} + R^{\alpha}_{\mu\rho\sigma;\nu} = 0 . \tag{3.126}$$

In addition to the symmetry relationships derived earlier, the Riemann tensor satisfies the differential equations above known as the *Bianchi identities*.

### 3.4.6  EINSTEIN TENSOR

Let us multiply the differential equations for the Bianchi identities (3.126) by $g^{\mu\nu}$, contract $\sigma$ with $\alpha$, and use the fact, already established, that the covariant derivatives of the metric tensor vanish:

$$\begin{aligned} 0 &= g^{\mu\nu}(R^{\alpha}_{\mu\nu\rho;\sigma} + R^{\alpha}_{\mu\sigma\nu;\rho} + R^{\alpha}_{\mu\rho\sigma;\nu}) \\ &= (g^{\mu\nu}R^{\alpha}_{\mu\nu\rho})_{;\alpha} + (g^{\mu\nu}R^{\alpha}_{\mu\alpha\nu})_{;\rho} + (g^{\mu\nu}R^{\alpha}_{\mu\rho\alpha})_{;\nu} . \end{aligned} \tag{3.127}$$

Examine each term in brackets using the Riemann tensor symmetries. The first term is

$$g^{\mu\nu}R^{\alpha}_{\mu\nu\rho} = g^{\mu\nu}g^{\alpha\beta}R_{\beta\mu\nu\rho} = g^{\mu\nu}g^{\alpha\beta}R_{\mu\beta\rho\nu} = g^{\alpha\beta}R^{\nu}_{\beta\rho\nu} = g^{\alpha\beta}R_{\beta\rho}$$
$$= R^{\alpha}_{\rho} .$$

The second term is

$$g^{\mu\nu} R^\alpha_{\mu\alpha\nu} = -g^{\mu\nu} R^\alpha_{\mu\nu\alpha} = -g^{\mu\nu} R_{\mu\nu} = -R\,.$$

The third term is

$$g^{\mu\nu} R^\alpha_{\mu\rho\alpha} = g^{\mu\nu} R_{\mu\rho} = R^\nu_\rho\,.$$

Now put these results back into their brackets with the covariant derivatives as indicated in (3.127) to obtain

$$0 = R^\alpha_{\rho;\alpha} - R_{;\rho} + R^\nu_{\rho;\nu} = 2R^\alpha_{\rho;\alpha} - R_{;\rho}\,.$$

Multiply by $g^{\mu\rho}$, and note that

$$g^{\mu\rho} R^\alpha_{\rho;\alpha} = (g^{\mu\rho} R^\alpha_\rho)_{;\alpha} = R^{\mu\alpha}_{\ \ ;\alpha} = R^{\mu\nu}_{\ \ ;\nu}$$
$$g^{\mu\rho} R_{;\rho} = g^{\mu\nu} R_{;\nu}$$

to arrive immediately at the vanishing divergence

$$(R^{\mu\nu} - \tfrac{1}{2} g^{\mu\nu} R)_{;\nu} = 0\,. \tag{3.128}$$

The object in the brackets is called the *Einstein curvature tensor*,

$$G^{\mu\nu} \equiv R^{\mu\nu} - \tfrac{1}{2} g^{\mu\nu} R\,. \tag{3.129}$$

The Einstein tensor is constructed from the Riemann curvature tensor and has an identically vanishing covariant divergence. It is symmetric and of second rank. Einstein was motivated to seek a tensor that contained no differentials of the $g^{\mu\nu}$ higher than the second—a tensor which was a linear homogeneous combination of terms linear in the second derivative or quadratic in the first (in analogy with Poisson's equation for the gravitational potential in Newton's theory:

$$\nabla^2 V = 4\pi\rho\,, \tag{3.130}$$

where $\rho$ is the mass density generating the field). For the expression of energy and momentum conservation, it is important that the divergence vanish. The energy-momentum tensor of matter accomplishes this and is of second rank.

## 3.5   Einstein's Field Equations

"The geometry of spacetime is not given; it is determined
by matter and its motion."[5]
W. Pauli, 1919

We know that other bodies will experience gravity in the vicinity of massive bodies. So mass is a source of gravity, and from the Special Theory of Relativity we must say in general that mass and energy are sources. We have just seen that a construction from the Riemann curvature tensor, namely, Einstein's tensor, has vanishing covariant divergence. We have three possibilities,

$$G^{\mu\nu} = 0 \,, \tag{3.131}$$

or

$$G^{\mu\nu} = kT^{\mu\nu} \,, \tag{3.132}$$

where $T^{\mu\nu}$ is a symmetric divergenceless tensor constructed from the mass-energy properties of the material medium, or

$$G^{\mu\nu} = kT^{\mu\nu} + \Lambda g^{\mu\nu} \,. \tag{3.133}$$

The constant $k$ will be found in connection with the Oppenheimer–Volkoff equations, discussed later, to be $8\pi$. The constant $\Lambda$ is the so-called *cosmological constant*. It was not present in the original theory and was added to obtain a static cosmology before it was known that the universe is expanding. Einstein regarded its numerical value as a matter to be settled by experiment—"The curvature constant [$\Lambda$] is, however, essentially determinable, and an increase in the precision of data derived from observations will enable us in the future to fix its sign and determine its value" [34].

It is apparent that the cosmological constant corresponds to a constant energy density $\Lambda/(8\pi)$ and a constant pressure of the same numerical value but of opposite sign. The cosmological constant is sometimes referred to as the vacuum energy density. In any case it is small; its value has been recently measured [35, 36]. (See also ref. [37] for further discussion.)

Its effect is indeed cosmological; stellar structure is unaffected by it. We need not consider the cosmological term further as regards stars. (However, as has been discovered only recently, on the cosmological scale it is of extreme importance, for it causes the expansion of the universe to accelerate after the gravitational attraction of photons and matter have diluted sufficiently. The acceleration era began about 2 billion years ago.

The first set of differential equations (3.131) are those that must be satisfied by the metric in empty space outside material bodies and energy concentrations. An example is the gravitational fields outside a star.

---

[5] Very importantly, the converse is also true.

The second set of differential equations (3.132) determine the gravitational fields $g^{\mu\nu}$ inside a spacetime region of mass-energy and in addition determine how the mass-energy is arranged by gravity. With appropriate $T^{\mu\nu}$ it would provide the equations of stellar structure. We have yet to fix the constant $k$. This can be done by looking to the weak field limit where the General Theory of Relativity should agree with Newton's well-tested, weak-field theory.

There are several remarkable notes we can make at this point. Einstein's field equations tell spacetime how to curve and mass-energy how to configure itself and how to move. Spacetime acts upon matter and in turn is acted upon by matter. This was Einstein's intuition and motivation in seeking a theory that placed spacetime and matter as co-determiners in nature. He was displeased with the Special Theory of Relativity as anything but a local theory, for it gave spacetime an absolute status.

Second, the Einstein field equations are nonlinear in the fields $g^{\mu\nu}$. (This can be verified by tracing back through the objects from which the Einstein tensor is constructed.) Nonlinearity means that the gravitational field interacts with itself. This is because the field carries energy, and mass-energy in any form is a source of gravity. The nonlinearity of the Einstein equations accounts for some of the extraordinary phenomena encountered in strong gravity, including black holes [38] and the reversal of the centrifugal force in their vicinity [39].

We have seen in (3.128) that the Einstein tensor has identically vanishing covariant divergence. Hence (3.132) requires of the matter tensor that

$$T^{\mu\nu}{}_{;\nu} = 0. \tag{3.134}$$

The corresponding equation in flat space is

$$T^{\mu\nu}{}_{,\nu} = 0. \tag{3.135}$$

Vanishing of the ordinary divergence of the energy-momentum tensor in the Special Theory of Relativity corresponds to the conservation of energy and momentum. However, (3.134) does not assure us of the constancy of any quantity in time. In fact (3.132) ensures that matter and the gravitational fields exchange energy, or in other words do work on each other, for it is the divergence of $G^{\mu\nu} - kT^{\mu\nu}$ that vanishes. So neither matter nor the gravitational field can by itself conserve energy in any sense. No contradiction exists with laboratory experiments performed on earth. Over the dimensions of a typical laboratory, spacetime is essentially flat, and nothing that could be done in a laboratory could possibly disturb this flatness in any perceptible way.

This brings us back to the comparison of the weak-field limit between Newton's theory and Einstein's. The inverse-square law of the force between massive objects is not required by the inner structure of Newton's theory. He could have postulated an inverse $\alpha$ law: $F \sim Mm/r^{\alpha}$. And then attempted to fit $\alpha$ to the astronomical data of the solar system. Depending on what weight was given to the precession of planets, one would have found a value of $\alpha$ close to two.

Einstein's theory does not possess the flexibility of Newton's in this regard. We saw in (3.72) that Einstein's theory, in the limit of a weak gravitational field, predicts precisely the inverse square law. In this sense, he could claim as his own all the successes of the Newtonian theory in explaining the motion of planets in the solar system. They were computed with the inverse-square law, there being no flexibility in the choice of the power in his theory.

Concerning the precession of planets, an isolated planet in orbit about the Sun under an inverse square law is an ellipse whose orientation is fixed in space. However the total precession of the orbit of Mercury is observed to be about 5600 seconds/century. Most of this is caused by the fact that an earthbound observer is not in an inertial frame far from the Sun. For example, suppose that Mercury did not orbit about the Sun, but instead held a fixed position. Nevertheless, from the earth it would be appear to move, sometimes to the left of the Sun, sometimes to the right, and alternately passing in front of and in back of the sun. Taking account of this correction to the apparent motion of Mercury due to the earth's own motion, the precession of Mercury is about 574 seconds/century. This value is about 43 seconds/century larger than the precession computed by Newtonian physics as due to the perturbation of the orbit by other planets, a small but disturbing discrepancy. An early triumph of Einstein was that he calculated, within the observational errors, the precise value of the excess precession. In Newton's theory only mass contributes to gravity, whereas in Einstein's theory the kinetic energy of the motion of the planets contributes as well.

## 3.6   Relativistic Stars

Einstein's field equations are completely general and simple in appearance. However, they are exceedingly complicated because of their nonlinear character and because spacetime and matter act upon each other. As already remarked, there is no prior geometry of spacetime. There are a few cases in which solutions can be found in closed form. One of the most important closed-form solutions is the Schwarzschild metric outside a static spherical star. Another is the Kerr metric outside a rotating black hole. Einstein's equations can also be solved numerically as the coupled differential equations for the interior structure of a spherical static star, which are called the Oppenheimer–Volkoff equations for stellar structure.

In this section we take up the important problem of deriving the equations that govern spacetime and the arrangement of matter in the case of relativistic spherical static stars. They are the basic equations that underlie the development of neutron star models. They also demonstrate the mathematical existence of Schwarzschild black holes. They can also be used to develop white dwarf models, though Newtonian gravity is a good approximation for these stars.

### 3.6.1  METRIC IN STATIC ISOTROPIC SPACETIME

We seek solutions to Einstein's field equations in static isotropic regions of spacetime such as would be encountered in the interior and exterior regions of static stars. Under these conditions the $g_{\mu\nu}$ are independent of time ($x^0 \equiv t$) and $g^{0m} = 0$. We choose spatial coordinates $x^1 = r$, $x^2 = \theta$, and $x^3 = \phi$. The most general form of the line element is then

$$d\tau^2 = U(r)\, dt^2 - V(r)\, dr^2 - W(r)r^2 (\, d\theta^2 + \sin^2\theta\, d\phi^2)\,. \qquad (3.136)$$

We may replace $r$ by any function of $r$ without disturbing the spherical symmetry. We do so in such a way that $W(r) \equiv 1$. Then we may write

$$d\tau^2 = e^{2\nu(r)}\, dt^2 - e^{2\lambda(r)}\, dr^2 - r^2\, d\theta^2 - r^2 \sin^2\theta\, d\phi^2\,, \qquad (3.137)$$

where $\lambda, \nu$ are functions only of $r$. Comparing with

$$d\tau^2 = g_{\mu\nu}\, dx^\mu\, dx^\nu,$$

we read off [6]

$$\begin{aligned}
g_{00} &= e^{2\nu(r)}, \quad g_{11} = -e^{2\lambda(r)}, \quad g_{22} = -r^2, \quad g_{33} = -r^2 \sin^2\theta, \\
g_{\mu\nu} &= g^{\mu\nu} = 0 \quad (\mu \neq \nu)\,.
\end{aligned} \qquad (3.138)$$

Hence, from $g_{\mu\nu} g^{\nu\rho} = \delta^\rho_\mu$, we have in this special case

$$g_{\mu\mu} = 1/g^{\mu\mu} \quad \text{(not summed)}\,. \qquad (3.139)$$

According to its definition as a contraction of the Riemann tensor, the Ricci tensor can be written

$$R_{\mu\nu} = \Gamma^\alpha_{\mu\alpha,\nu} - \Gamma^\alpha_{\mu\nu,\alpha} - \Gamma^\alpha_{\mu\nu}\Gamma^\beta_{\alpha\beta} + \Gamma^\alpha_{\mu\beta}\Gamma^\beta_{\nu\alpha}\,. \qquad (3.140)$$

We can derive the nonvanishing affine connections (3.46), which are symmetric in their lower indices, from the metric tensor whose general form for static isotropic regions was derived above:

$$\begin{aligned}
\Gamma^1_{00} &= \nu'\, e^{2(\nu-\lambda)}\,, & \Gamma^0_{10} &= \nu'\,, \\
\Gamma^1_{11} &= \lambda'\,, & \Gamma^2_{12} &= \Gamma^3_{13} = 1/r\,, \\
\Gamma^1_{22} &= -re^{-2\lambda}\,, & \Gamma^3_{23} &= \cot\theta\,, \\
\Gamma^1_{33} &= -r\sin^2\theta\, e^{-2\lambda}\,, & \Gamma^2_{33} &= -\sin\theta\cos\theta\,.
\end{aligned} \qquad (3.141)$$

---

[6]Note that in (3.137) some authors use the opposite signs for time and space components, and some use the functions $\nu$, $\lambda$ but without the factor 2, or use different notation altogether for the metric. Great care has to be exercised in using results from different sources.

The primes denote differentiation with respect to $r$. Hence, for static isotropic spacetime

$$
\begin{aligned}
R_{00} &= \left(-\nu'' + \lambda'\nu' - \nu'^2 - \frac{2\nu'}{r}\right)e^{2(\nu-\lambda)}\,, \\
R_{11} &= \nu'' - \lambda'\nu' + \nu'^2 - \frac{2\lambda'}{r}\,, \\
R_{22} &= (1 + r\nu' - r\lambda')e^{-2\lambda} - 1\,, \\
R_{33} &= R_{22}\sin^2\theta\,.
\end{aligned} \tag{3.142}
$$

## 3.6.2   THE SCHWARZSCHILD SOLUTION

In the empty space outside a static star Einstein's equation is $G_{\mu\nu} = 0$, or equivalently

$$
R_{\mu\nu} = \tfrac{1}{2}g_{\mu\nu}R\,. \tag{3.143}
$$

Multiply by $g^{\alpha\mu}$, and sum on the dummy index to find

$$
R^\alpha_\nu = \tfrac{1}{2}\delta^\alpha_\nu R\,. \tag{3.144}
$$

Contract by setting $\alpha = \nu$, and sum to get

$$
R = 2R \quad \Longrightarrow \quad R = 0\,. \tag{3.145}
$$

Hence, the vanishing of Einstein's tensor implies

$$
G_{\mu\nu} = 0 \quad \Longrightarrow \quad R = 0,\quad R_{\mu\nu} = 0\,. \tag{3.146}
$$

In empty space, Einstein's equation is equivalent to the vanishing of the Ricci tensor or, equivalently, the scalar curvature. Its form for static isotropic spacetime was worked out in the previous section.

From the vanishing of $R_{00}, R_{11}$ we find that

$$
\lambda' + \nu' = 0\,. \tag{3.147}
$$

(Do not confuse $\nu$ and $\lambda$ when used to denote indices and when used to denote the metric functions as in the above equation.) For large $r$, space must be unaffected by the star and therefore flat so that $\lambda$ and $\nu$ tend to zero; therefore

$$
\lambda + \nu = 0\,. \tag{3.148}
$$

Using these results in $R_{22} = 0$, we find that

$$
(1 + 2r\nu')e^{2\nu} = 1\,. \tag{3.149}
$$

This condition integrates to

$$
g_{00} \equiv e^{2\nu} = 1 - \frac{2GM}{r} \qquad (r > R)\,, \tag{3.150}
$$

where $M$ is the constant of integration, and we introduced Newton's constant. Having studied the Newtonian approximation of Einstein's theory (3.72), one identifies $M$ with the mass of the star. From the foregoing results,

$$g_{11} = -e^{2\lambda} = -e^{-2\nu} = -\left(1 - \frac{2GM}{r}\right)^{-1} \quad (r > R). \tag{3.151}$$

This completes the derivation of the *Schwarzschild solution* of 1916 of Einstein's equations outside a spherical static star. It was the first exact solution found for Einstein's equations. The proper time is

$$d\tau^2 = \left(1 - \frac{2GM}{r}\right)dt^2 - \left(1 - \frac{2GM}{r}\right)^{-1}dr^2$$
$$-r^2 d\theta^2 - r^2 \sin^2\theta \, d\phi^2, \quad (r > R) \tag{3.152}$$

where $R$, in this context, denotes the radius of the star.

Let us summarize the Schwarzschild solution found above for $r > R$:

$$g_{00}(r) = e^{2\nu(r)} = \left(1 - \frac{2GM}{r}\right),$$
$$g_{11}(r) = -e^{2\lambda(r)} = -\left(1 - \frac{2GM}{r}\right)^{-1},$$
$$g_{22}(r) = -r^2, \quad g_{33}(r, \theta) = -r^2 \sin^2\theta. \tag{3.153}$$

Notice that the Schwarzschild metric is singular at the radius

$$r = r_S \equiv 2GM/c^2, \tag{3.154}$$

where we have reinserted $c$ which by our usual convention we take as unity.

This does not mean that spacetime itself is singular at that radius, but only that this particular metric is. Other nonsingular metrics have been found, in particular, the Kruskal–Szerkeres metric [40, 41]. However, further analysis shows that if $r_S$ lies outside the star where the Schwarzschild solution holds, then it is a black hole—no particle or even light can leave the region $r < r_S$. This radius, $r_S$, is called the *Schwarzschild radius* or *singularity* or *horizon*. But because the above metric holds only outside the star, $r_S$ has no special significance if it is smaller than the radius of the star. For then a different metric holds inside the star which does not possess a singularity. We come to this solution shortly.

### 3.6.3   RIEMANN TENSOR OUTSIDE A SCHWARZSCHILD STAR

If spacetime is flat, then the Riemann curvature tensor vanishes (Section 3.4.3). We are now prepared to address the converse (albeit not rigorously): if spacetime is curved, some components of the Riemann tensor are finite (which components, of course, will depend upon how convoluted spacetime is).

The metric tensor and, indeed, the affine connection for the empty space outside a massive body were computed in the preceding section. We have seen in Section 3.1.3 that massive bodies curve spacetime. So we know that the Schwarzschild metric tensor refers to curved spacetime. Referring to the definition of the Riemann tensor (3.114) and the specific form that the affine connection takes for a static spherical star (3.141), we can compute

$$R_{010}^1 = \left(\nu'' + 2\nu'^2 - \nu'\lambda'\right)e^{2(\nu-\lambda)}. \tag{3.155}$$

Thus we exhibit at least one nonvanishing component of the Riemann tensor in the curved spacetime outside a Schwarzschild star. This suggests that Riemann is not identically zero in curved spacetime. An actual proof that if Riemann is finite then spacetime is curved requires the formulation of parallel transport, which we do not take up here. We declare, without rigorous proof, that the Riemann tensor vanishes if and only if spacetime is flat. Notice that, far from an isolated star where spacetime approaches flatness, Riemann approaches zero as it should.

### 3.6.4   ENERGY-MOMENTUM TENSOR OF MATTER

From the success of Newtonian physics in describing celestial mechanics and other weak gravitational field phenomena, we know that mass is a source of gravity. From the experimental verifications of the Special Theory of Relativity, we know that all forms of energy are equivalent and must contribute equally as sources of gravity. Normally, of course, it is mass that dominates, and the average mass density in the solar system and in the universe is very small; that is why Newtonian physics is so accurate under the typical conditions mentioned above.

An essential aspect of Einstein's curvature tensor is that it automatically has vanishing covariant divergence (3.128). It is also a symmetric second-rank tensor. Accordingly, mass-energy—the source of the gravitational field—must be incorporated into a divergenceless, symmetric, second-rank tensor in flat space. As a tensor, it can be transcribed immediately to its form in an arbitrary spacetime frame by the general covariance principle. Such a tensor is the energy-momentum tensor.

In the book *Compact Stars* [4] specific theories of dense matter are described from which we are able to explicitly construct the energy-momentum tensor of the theory. Here we are interested in the general form such a tensor takes. Frequently, matter may be regarded as a perfect fluid. The fluid velocity is assumed to vary continuously from point to point. The perfect fluid energy-momentum tensor in the Special Theory of Relativity can be expressed in terms of the local values of the pressure $p$ and energy density $\epsilon$ as in (2.54). The General Relativistic energy-momentum tensor can be written immediately using the Principle of General Covariance spelled out on page 42:

$$T^{\mu\nu} = -pg^{\mu\nu} + (p+\epsilon)u^\mu u^\nu \,,$$
$$g_{\mu\nu}u^\mu u^\nu = u_\nu u^\nu = 1 \,. \tag{3.156}$$

In the above equations, $u^\mu$ is the local fluid four-velocity (see 2.47)

$$u^\mu = dx^\mu/d\tau \,. \tag{3.157}$$

The pressure and total energy density (including mass) are related by the equation of state of matter, frequently written in either form

$$p = p(\epsilon) \quad \text{or} \quad \epsilon = \epsilon(p) \tag{3.158}$$

where $p$ and $\epsilon$ are the pressure and energy density (including mass) in the local rest-frame of the fluid. In the next section we shall see how the equations for stellar structure involve these quantities and this relationship.

### 3.6.5   THE OPPENHEIMER–VOLKOFF EQUATIONS

We are now prepared to derive the differential equations for the structure of a static, spherically symmetric, relativistic star. For the region outside a star, we found that the vanishing of the Einstein tensor was equivalent to the vanishing of the Ricci tensor or the scalar curvature. This is not the case for the interior of the star. We need both the Ricci tensor and scalar curvature to construct the Einstein tensor. The general form of the metric for a static isotropic spacetime was obtained in (3.138). From Section 3.6.1 we find the scalar curvature,

$$
\begin{aligned}
R &= g^{\mu\nu}R_{\mu\nu} = e^{-2\nu}R_{00} - e^{-2\lambda}R_{11} - \frac{2}{r^2}R_{22} \\
&= e^{-2\lambda}\left\{ -2\nu'' + 2\lambda'\nu' - 2\nu'^2 - \frac{2}{r^2} + 4\frac{\lambda'}{r} - 4\frac{\nu'}{r} \right\} \\
&\quad + \frac{2}{r^2} \,.
\end{aligned}
\tag{3.159}
$$

It is more convenient to work with mixed tensors. For example,

$$G_0{}^0 = R_0{}^0 - \tfrac{1}{2}R \tag{3.160}$$

is obtained with the results of Section 3.6.1 for a static isotropic field, namely,

$$g_0^0 = g_{0\nu}g^{0\nu} = g_{00}g^{00} = 1 \,. \tag{3.161}$$

So using results obtained earlier in this section we can find that the components of the Einstein tensor are

$$r^2 G_0{}^0 \equiv e^{-2\lambda}(1 - 2r\lambda') - 1 = -\frac{d}{dr}[r(1 - e^{-2\lambda})] \,,$$

$$r^2G_1{}^1 \equiv e^{-2\lambda}(1 + 2r\nu') - 1,$$

$$G_2{}^2 \equiv e^{-2\lambda}\left(\nu'' + \nu'^2 - \lambda'\nu' + \frac{\nu' - \lambda'}{r}\right),$$

$$G_3{}^3 = G_2{}^2. \tag{3.162}$$

Because of the assumption that the star is static, the three-velocity of every fluid element is zero, so

$$u^\mu = 0 \quad (\mu \neq 0), \qquad u^0 = 1/\sqrt{g_{00}}, \tag{3.163}$$

according to (3.156). The energy-momentum tensor expressed as a mixed tensor, we have the nonzero components in the present metric,

$$T_0{}^0 = \epsilon, \quad T_\mu{}^\mu = -p \quad (\mu \neq 0). \tag{3.164}$$

So the (00) component of the Einstein equations gives

$$r^2G_0{}^0 = -\frac{d}{dr}\{r(1 - e^{-2\lambda(r)})\} = kr^2T_0{}^0 = kr^2\epsilon(r). \tag{3.165}$$

This can be integrated immediately to yield

$$e^{-2\lambda(r)} = 1 + \frac{k}{r}\int_0^r \epsilon(r)r^2 dr. \tag{3.166}$$

Let us define

$$M(r) \equiv 4\pi \int_0^r \epsilon(r)r^2 dr, \tag{3.167}$$

and let $R$ denote the radius of the star, the radial coordinate exterior to which the pressure vanishes. Zero pressure defines the edge of the star because zero pressure can support no material against the gravitational attraction from within. Denote the corresponding value of $M(R)$ by

$$M \equiv M(R). \tag{3.168}$$

Now comparing (3.72, 3.150, 3.151) we see that, to obtain agreement with the Newtonian limit, we must choose

$$k = -8\pi G \tag{3.169}$$

and interpret $M$ as the *gravitational mass* of the star. Therefore, $M(r)$ is referred to as the *included mass* within the coordinate $r$. So Einstein's field equations can now be written

$$G^{\mu\nu} = -8\pi G T^{\mu\nu}. \tag{3.170}$$

From the above, we have found so far that

$$g_{11}(r) = -e^{2\lambda(r)} = -\left(1 - \frac{2GM(r)}{r}\right)^{-1}, \tag{3.171}$$

which agrees with (3.151), but now we see that $g_{11}(r)$ has the same form inside and outside the star although it is the included mass $M(r)$, not the total mass, that appears in the interior solution.

Having learned the constant of proportionality in Einstein's equations (3.170), let us now write out the field equations for a spherically symmetric static star, including the one we have already solved. In passing we note that our solution gives a relationship between the included mass $M(r)$ at any radial coordinate and the metric function $g_{11}(r)$ or $\lambda(r)$, but we have yet to learn how to compute one or the other. The differential equations from (3.162) are

$$G_0^{\ 0} \equiv e^{-2\lambda}\left(\frac{1}{r^2} - \frac{2\lambda'}{r}\right) - \frac{1}{r^2} = -8\pi G\epsilon(r),  \tag{3.172}$$

$$G_1^{\ 1} \equiv e^{-2\lambda}\left(\frac{1}{r^2} + \frac{2\nu'}{r}\right) - \frac{1}{r^2} = 8\pi G p(r),  \tag{3.173}$$

$$G_2^{\ 2} \equiv e^{-2\lambda}\left(\nu'' + \nu'^2 - \lambda'\nu' + \frac{\nu' - \lambda'}{r}\right) = 8\pi G p(r),  \tag{3.174}$$

$$G_3^{\ 3} = G_2^{\ 2} = 8\pi G p(r).  \tag{3.175}$$

The last equation contains no information additional to that provided by those preceding it.

To simplify notation, choose units so that $G = c = 1$. Solve (3.172) to find

$$-2r\lambda' = (1 - 8\pi r^2\epsilon)e^{2\lambda} - 1  \tag{3.176}$$

and (3.173) to find

$$2r\nu' = (1 + 8\pi r^2 p)e^{2\lambda} - 1.  \tag{3.177}$$

Take the derivative of (3.177) and then multiply by $r$:

$$2r\nu' + 2r^2\nu'' = \left[2r\lambda'(1 + 8\pi r^2 p) + (16\pi r^2 p + 8\pi r^3 p')\right]e^{2\lambda}.$$

Solve for $\nu''$ using (3.177, 3.176):

$$\begin{aligned} 2r^2\nu'' = {} & 1 + (16\pi r^2 p + 8\pi r^3 p')e^{2\lambda} \\ & -(1 + 8\pi r^2 p)(1 - 8\pi r^2\epsilon)e^{4\lambda}. \end{aligned}  \tag{3.178}$$

Square (3.177) to obtain the result

$$2r^2\nu'^2 = \tfrac{1}{2}(1 + 8\pi r^2 p)^2 e^{4\lambda} - (1 + 8\pi r^2 p)e^{2\lambda} + \tfrac{1}{2}.  \tag{3.179}$$

The last four numbered equations provide expressions for $\lambda', \nu', \nu''$, and $\nu'^2$ in terms of $p$, $p'$, $\epsilon$, and $e^{2\lambda}$ the latter of which, according to (3.171), can be expressed in terms of the included mass. Therefore the metric can be eliminated altogether by substitution of the above results into the remaining field equation (3.174). After a number of cancellations, we emerge with the result

$$\frac{dp}{dr} = -\frac{[p(r) + \epsilon(r)]\,[M(r) + 4\pi r^3 p(r)]}{r\,[r - 2M(r)]}. \tag{3.180}$$

This and equation (3.167) represent the reduction of Einstein's equations for the interior of a spherical, static, relativistic star. These equations are frequently referred to as the Oppenheimer–Volkoff equations. The stars they describe—static and spherically symmetric—are sometimes referred to as Schwarzschild stars. Notice that this metric becomes singular at

$$r = r_S = 2GM/c^2, \tag{3.181}$$

which is called the Schwarzschild radius (with $c$ and $G$ reinserted). The singularity in this metric is a feature of this metric, and there are other metrics in which it is seen that no singularity exists across the boundary of a black hole.

To see the correspondence with Newton's equations for stellar structure, rewrite the above as,

$$r^2\frac{dp}{dr} = -\frac{\epsilon(r)M(r)[1 + p(r)/\epsilon(r)][1 + 4\pi r^3 p(r)/M(r)]}{[1 - 2M(r)/r]} \tag{3.182}$$

The three terms in square brackets are the General Relativistic corrections to Newton.

Given an equation of state (3.158), the stellar structure equations (3.167) and (3.180) can be solved simultaneously for the radial distribution of pressure, $p(r)$, and hence for the distribution of mass-energy density $\epsilon(r)$. Moreover, in any detailed theory of dense matter, the baryon and lepton populations are obtained as a function of density; hence the distribution of particle populations in a star can be found coincident with a solution of the Oppenheimer–Volkoff equations.

It may seem curious that the expression (3.167) for mass has precisely the same form as one would write in nonrelativistic physics for the mass whose distribution is given by $\epsilon(r)$. How can this be, inasmuch as we know that spacetime is curved by mass and mass in turn is moved and arranged by spacetime in accord with Einstein's equations? The answer is that (3.167) is not a prescription for computing the total mass of an arbitrary distribution $\epsilon(r)$. There are *no* arbitrary distributions in gravity; rather $\epsilon(r)$ is precisely prescribed by another of Einstein's equations (3.180) together with the equation of state, which relates $\epsilon$ and $p$. As such, $M$ comprises the mass of the star and its gravitational field. Because of the mutual interaction of mass-energy and spacetime, there is no meaning to the question "What is the mass of the star?" in isolation from the field energy. That is why we refer to $M$ as the gravitational mass or the mass-energy of the star. It is the only type of mass that enters Einstein's theory and is the only stellar mass to which we will refer in this book. Therefore, we shall generally refer to a star's mass as simply the mass without the adjective "gravitational". Sometimes a so-called proper mass is defined. It appears nowhere in Einstein's equations and is an artifact. It is not measurable.

It does make sense to inquire about the mass of the totality of nucleons in a star if they were dispersed to infinity. This mass is referred to as the baryon mass. The difference between gravitational mass and baryon mass, if negative, is the gravitational binding of the star. As we shall find, the gravitational binding is of the order of 100 MeV per nucleon in stars near the mass limit as compared to 10 MeV binding by the strong force in nuclei.

Notice that, according to (3.180), the pressure is a monotonic decreasing function from the inside of the star to its edge because all the factors in (3.180) are positive, leaving the explicit negative sign. This makes sense. Any region is weighted down by all that lies above. We have assumed that the denominator in (3.180) is positive. Overall this is true of the earth, the Sun, and a neutron star. In fact, $2M/R < 8/9$ for any static stable star. It can also be shown that $2M(r)/r < 1$ for all regions of a stable star [42]; so indeed we are justified in taking the denominator in (3.180) as positive.

In (3.152) we saw a singularity in the Schwarzschild solution if a star lies within $r = 2M$. Such stars are highly relativistic objects called black holes. No light or particle can escape from within their Schwarzschild radius. A luminous star is highly nonrelativistic. A neutron star is relativistic. Newtonian gravity would not produce the same results as General Relativity. This fact is clear, given that $2M$ can be as large as $\frac{8}{9}R$ for a neutron star, which makes the denominator of (3.180) a large correction (as much as 9 instead of 1).

We already have an expression for the radial metric function both inside and outside a star. It is sometimes useful to have the time metric function $g_{00}$. No general expression for the solution can be obtained, as for $g_{11}$, (3.171). However using the latter in (3.177) we obtain a differential equation,

$$\frac{d\nu}{dr} = \frac{M(r) + 4\pi r^3 p(r)}{r\left[r - 2M(r)\right]}. \tag{3.183}$$

The solution must match the exterior solution (3.151). This is easily accomplished. If $\nu(r)$ is a solution, we can add any constant to it and still have a solution. We obtain the correct condition at $R$ if we make the change

$$\nu(r) \longrightarrow \nu(r) - \nu(R) + \tfrac{1}{2}\ln\left(1 - \frac{2M}{R}\right), \qquad r \leq R. \tag{3.184}$$

We can start the integration at $r = 0$ with any convenient value of $\nu(0)$, say zero.

Alternately, once the OV equations have been solved so that $p(r)$ and hence $\epsilon(r)$ are known, one can find $\nu(r)$ by integration of

$$\frac{d\nu}{dr} = -\frac{1}{p+\epsilon}\frac{dp}{dr}, \qquad (3.185)$$

namely,

$$\nu(r) = -\int_0^r \frac{1}{p+\epsilon}\frac{dp}{dr} + \text{constant}, \qquad \nu(\infty) = 0. \qquad (3.186)$$

The Oppenheimer–Volkoff equations can be integrated from the origin with the initial conditions $M(0) = 0$ and an arbitrary value for the central energy density $\epsilon(0)$, until the pressure $p(r)$ becomes zero at, say $R$. Because zero pressure can support no overlying matter against the gravitational attraction, $R$ defines the gravitational radius of the star and $M(R)$ its gravitational mass. For the given equation of state, there is a unique relationship between the mass and central density $\epsilon(0)$. So for each possible equation of state, there is a unique family of stars, parameterized by, say, the central density or the central pressure. Such a family is often referred to as the *single parameter sequence* of stars corresponding to the given equation of state.

## 3.6.6   GRAVITATIONAL COLLAPSE AND LIMITING MASS

In Newtonian physics, mass alone generates gravity. In the Special Theory of Relativity mass is equivalent to energy, so in General Relativity all forms of energy contribute to gravity. It is surprising that pressure also plays a most consequential role in the structure of relativistic stars beyond the role it plays in Newtonian gravity. Pressure supports stars against gravity, but surprisingly, it ultimately assures the gravitational collapse of relativistic stars whose mass lies above a certain limit.

Pressure appears together with energy density in determining the monotonic decrease of pressure (3.180) in a relativistic star. Gravity acts to compress the material of the star. As it does so, the pressure of the material is increased toward the center. But inasmuch as pressure appears on the right side of the equation, this increase serves to further enhance the grasp of gravity on the material. Therefore, for stars of increasing mass, for which the supporting pressure must correspondingly increase, the pressure gradient (which is negative) is increased in magnitude, making the radius of the star smaller because its edge necessarily occurs at $p = 0$. As a consequence, if the mass of a relativistic star exceeds a critical value, there is no escape from gravitational collapse to a black hole [43]. Whatever the equation of state, the one-parameter sequence of stable configurations belonging to that equation of state is terminated by a maximum-mass compact star. The mass of this star is referred to as the *mass limit* or *limiting mass* of the sequence. We see examples of this in the next chapter.

# 3.7   Action Principle in Gravity

We arrived at Einstein's equations by noting the vanishing divergence of the Einstein curvature tensor and equating it to the energy-momentum tensor of matter as the source of the gravitational field. We did not comment on how the energy-momentum tensor might be obtained. In general, this tensor is not given but must be calculated from a theory of matter. In what frame should the theory be solved? Evidently in the general frame of the gravitational field. But this is an entirely different problem than is normally solved in many-body theory.

We are accustomed to solving problems in nuclear and particle theory in flat spacetime (or even flat space) in which the constant Minkowski metric $\eta_{\mu\nu}$ appears, not a general and as yet unspecified field $g_{\mu\nu}(x)$. A tacit assumption is made in passing from the energy-momentum tensor in a Lorentz frame (2.54) to its form (3.156) in a general frame by means of the principle of general covariance, as was done in deriving the Oppenheimer–Volkoff equations of stellar structure. The local region over which Lorentz frames extend is assumed to be sufficiently large that the equations of motion of the matter fields can be solved in a Lorentz frame, that is, in the absence of gravity, and the corresponding energy-momentum tensor constructed from the solution for such a region.

As we shall see in the next chapter, the local inertial frames in the gravitational field of neutron stars (and therefore for the less dense white dwarfs and all other stars) are actually sufficiently extensive that the matter from which they are constituted can be described by theories in flat Minkowski spacetime. We shall refer to such a situation as a partial decoupling of matter from gravity. In other words, the equations of motion for the matter and radiation fields can be solved in Minkowski spacetime. The solutions will provide the means of calculating the energy density and pressure of matter $\epsilon$ and $p$ throughout the star. But the general metric functions of gravity $g_{\mu\nu}(x)$ reappear on the right side of Einstein's field equations in the energy-momentum tensor, (3.156), when referred to a general frame in accord with the principle of general covariance. Therefore the gravitational fields $g_{\mu\nu}(x)$ still appear on both sides of Einstein's field equations, and matter in bulk shapes spacetime just as spacetime shapes and moves matter in bulk. However, the local structure of matter is determined only by the equations of motion in Minkowski spacetime.

There are, theoretically, conceivable situations where the partial decoupling just described may not hold. In that case the equations of motion themselves contain, not the Minkowski metric tensor (a diagonal tensor with constant elements), but the general, spacetime-dependent, metric tensor. This is the fully coupled problem and obviously would be enormously difficult to solve. While we do not encounter this situation in this book (see as an example where strong coupling is used, Refs. [44, 45]), nonetheless it is worth seeing in symbolic form what the fully coupled problem looks like. The expectation that the stress-energy tensor should be obtained in general from a theory of matter by solving the field equations of the theory in a general gravitational field will be verified. It is also interesting to see Einstein's equations emerge from a variational principle.

We employ the gravitational action principle. As in all cases, the Lagrangian of gravity ought to be a scalar. We have encountered the Ricci scalar curvature $R = g^{\mu\nu}R_{\mu\nu}$, and from it, as Hilbert did, the Lagrangian density can be formed (with a prefactor that can be known only in hindsight):

$$\mathcal{L}_g = -\frac{1}{16\pi G}R\sqrt{-g}. \tag{3.187}$$

Here $G$ is Newton's constant and $g$ is the determinant of the metric $g_{\mu\nu}$, which is negative for our choice of signature for the metric. (Recall, for example, the Minkowski metric, or the Schwarzschild metric.) We also define the Lagrangian density

$$\mathcal{L}_m = L_m\sqrt{-g} \tag{3.188}$$

from the Lagrangian $L_m$ of the matter and radiation fields $\phi$. The total action is

$$I = \int (\mathcal{L}_g + \mathcal{L}_m)\,d^4x. \tag{3.189}$$

The coupled field equations for the matter and metric functions emerge as the conditions that yield vanishing variation of the action with respect to all the fields—the gravitational fields described by $g_{\mu\nu}$ and matter fields described by $\phi$. The manipulations are quite tedious and are relegated to the next section. The field equations obtained are

$$\frac{\partial L_m}{\partial \phi} - \partial_\mu \frac{\partial L_m}{\partial(\partial_\mu \phi)} = 0, \tag{3.190}$$

$$G^{\mu\nu} = -8\pi G T^{\mu\nu}, \tag{3.191}$$

where $G^{\mu\nu} \equiv R^{\mu\nu} - \frac{1}{2}g^{\mu\nu}R$ is the Einstein tensor(3.129). The first of the field equations reduces to the familiar Euler–Lagrange equations in the limit of weak gravitational fields,

$$g_{\mu\nu} \to \eta_{\mu\nu} \quad (\text{ weak gravity }).$$

One encounters the Euler–Lagrange equations in studying theories of dense nuclear matter [4]. The second are Einstein's field equations (3.170). The two are coupled through the metric tensor.

The matter-radiation energy-momentum tensor that emerges from the variational principle is given by

$$T^{\mu\nu} \equiv -g^{\mu\nu}L_m + 2\frac{\partial L_m}{\partial g_{\mu\nu}}. \tag{3.192}$$

The second term is

$$2\frac{\partial L_m}{\partial g_{\mu\nu}} = 2\frac{\partial L_m}{\partial(\partial_\alpha\phi)}\frac{1}{2}\frac{\partial(g_{\alpha\beta}\partial^\beta\phi)}{\partial g_{\mu\nu}} = \frac{\partial L_m}{\partial(\partial_\mu\phi)}\partial^\nu\phi .$$

Combining these results yields the canonical form of the energy-momentum tensor in field theory (for example, see Ref. [10]) except that the Minkowski tensor is replaced by the general metric of gravity. Thus we have

$$T^{\mu\nu} = -g^{\mu\nu}L_m + \sum_\phi \frac{\partial L_m}{\partial(\partial_\mu\phi)}g^{\nu\alpha}\partial_\alpha\phi , \qquad (3.193)$$

where the sum is over the various fields $\phi$ in $L_m$. In this way we see how the equations couple all matter and gravitational fields, $\phi(x), ..., g^{\mu\nu}(x)$ in the general case.

## 3.7.1  DERIVATIONS

We write down most of the steps in deriving the Einstein field and matter-radiation equations from the variation of the action. The gravitational and matter fields will be subjected to arbitrary variations except the values and first derivatives will be kept constant on the boundaries. We concentrate on the gravitational part because that is the hardest. First, from (3.19) and (3.49) we readily obtain by differentiation,

$$g^{\alpha\mu}{}_{,\sigma}\,g_{\mu\nu} + g^{\alpha\mu}\left(\Gamma_{\nu\mu\sigma} + \Gamma_{\mu\nu\sigma}\right) = 0$$

Multiply by $g^{\beta\nu}$ and sum on $\nu$ to find

$$g^{\alpha\beta}{}_{,\sigma} = -g^{\alpha\mu}\Gamma^\beta_{\mu\sigma} - g^{\beta\nu}\Gamma^\alpha_{\nu\sigma} . \qquad (3.194)$$

Next, evaluate

$$\begin{aligned}
(g^{\mu\nu}\sqrt{-g})_{,\sigma} &= g^{\mu\nu}{}_{,\sigma}\sqrt{-g} + g^{\mu\nu}(\sqrt{-g})_{,\sigma} \\
&= -g^{\mu\rho}\Gamma^\nu_{\rho\sigma}\sqrt{-g} - g^{\nu\rho}\Gamma^\mu_{\rho\sigma}\sqrt{-g} + g^{\mu\nu}(\sqrt{-g})_{,\sigma} .
\end{aligned}$$

Use (3.102) to find

$$(g^{\mu\nu}\sqrt{-g})_{,\sigma} = \left(-g^{\mu\rho}\Gamma^\nu_{\rho\sigma} - g^{\nu\rho}\Gamma^\mu_{\rho\sigma} + g^{\mu\nu}\Gamma^\rho_{\sigma\rho}\right)\sqrt{-g} . \qquad (3.195)$$

Now set $\sigma = \nu$ and contract. After a cancellation of two terms, find,

$$(g^{\mu\nu}\sqrt{-g})_{,\nu} = -g^{\nu\alpha}\Gamma^\mu_{\alpha\nu}\sqrt{-g} . \qquad (3.196)$$

The above results can now be employed to rewrite the gravitational action (where, as usual, we set $G = c = 1$ whenever convenient);

$$16\pi I_g = \int R\sqrt{-g}\,d^4x . \qquad (3.197)$$

From (3.122) define

$$L \equiv g^{\mu\nu} \left( \Gamma^{\sigma}_{\mu\nu} \Gamma^{\rho}_{\sigma\rho} - \Gamma^{\rho}_{\mu\sigma} \Gamma^{\sigma}_{\nu\rho} \right)$$
$$M \equiv g^{\mu\nu} \left( \Gamma^{\sigma}_{\mu\sigma,\nu} - \Gamma^{\sigma}_{\mu\nu,\sigma} \right) . \tag{3.198}$$

With these definitions the scalar curvature becomes

$$R = M - L . \tag{3.199}$$

Use (3.195) and (3.196) to find

$$M\sqrt{-g} = (g^{\mu\nu} \Gamma^{\sigma}_{\mu\sigma} \sqrt{-g})_{,\nu} - (g^{\mu\nu} \Gamma^{\sigma}_{\mu\nu} \sqrt{-g})_{,\sigma}$$
$$-(g^{\mu\nu} \sqrt{-g})_{,\nu} \Gamma^{\sigma}_{\mu\sigma} + (g^{\mu\nu} \sqrt{-g})_{,\sigma} \Gamma^{\sigma}_{\mu\nu} . \tag{3.200}$$

The first two terms are perfect differentials and so contribute nothing under the integral because of the vanishing of the fields and their derivatives on the boundaries. After some manipulation, the remaining two terms are found to be $2L\sqrt{-g}$. Consequently,

$$16\pi I_g = \int L\sqrt{-g}\, d^4x$$
$$= \int g^{\mu\nu} \left( \Gamma^{\sigma}_{\mu\nu} \Gamma^{\rho}_{\sigma\rho} - \Gamma^{\rho}_{\mu\sigma} \Gamma^{\sigma}_{\nu\rho} \right) \sqrt{-g}\, d^4x . \tag{3.201}$$

Now evaluate the variation, examining separately the two terms in the integrand. Use (3.102) and (3.196) to rewrite the variation of the first term;

$$\delta \left( g^{\mu\nu} \Gamma^{\sigma}_{\mu\nu} \Gamma^{\rho}_{\sigma\rho} \sqrt{-g} \right) =$$
$$\Gamma^{\alpha}_{\mu\nu} \delta(g^{\mu\nu}(\sqrt{-g})_{,\alpha}) - \Gamma^{\beta}_{\alpha\beta} \delta(g^{\alpha\nu} \sqrt{-g})_{,\nu} - \Gamma^{\beta}_{\alpha\beta} \Gamma^{\alpha}_{\mu\nu} \delta(g^{\mu\nu} \sqrt{-g}) .$$

Use (3.194) to develop the variation of the second term in (3.201) and find

$$\delta(g^{\mu\nu} \Gamma^{\rho}_{\mu\sigma} \Gamma^{\sigma}_{\nu\rho} \sqrt{-g}) = -\Gamma^{\alpha}_{\nu\beta} \delta(g^{\nu\beta}{}_{,\sigma} \sqrt{-g}) - \Gamma^{\beta}_{\mu\alpha} \Gamma^{\alpha}_{\nu\beta} \delta(g^{\mu\nu} \sqrt{-g}) .$$

Assemble the above two results and introduce the perfect differentials with compensating terms to form the variation of the integrand of (3.201);

$$\delta \mathcal{L}_g = [\Gamma^{\alpha}_{\mu\nu} \delta(g^{\mu\nu} \sqrt{-g})]_{,\alpha} - [\Gamma^{\beta}_{\alpha\beta} \delta(g^{\alpha\nu} \sqrt{-g})]_{,\nu}$$
$$+(-\Gamma^{\alpha}_{\mu\nu,\alpha} + \Gamma^{\beta}_{\mu\beta,\nu} + \Gamma^{\beta}_{\mu\alpha} \Gamma^{\alpha}_{\nu\beta} - \Gamma^{\beta}_{\alpha\beta} \Gamma^{\alpha}_{\mu\nu}) \delta(g^{\mu\nu} \sqrt{-g}) .$$

The first two terms are perfect differentials and yield zero because the variations vanish on the boundary. The remaining bracket is the Ricci tensor (3.122). Therefore, we have for the variation of the gravitational action

$$\delta I_g = \frac{1}{16\pi} \int R_{\mu\nu} \delta(g^{\mu\nu} \sqrt{-g})\, d^4x . \tag{3.202}$$

Next, again from (3.19) deduce that

$$\delta g^{\alpha\nu} = -g^{\mu\alpha}g^{\nu\beta}\delta g_{\alpha\beta}. \tag{3.203}$$

Also, from (3.101) find

$$2\sqrt{-g}(\sqrt{-g})_{,\alpha} = (\sqrt{-g}\sqrt{-g})_{,\alpha} = g_{,\alpha} = gg^{\mu\nu}g_{\mu\nu,\alpha}.$$

Consequently,

$$(\sqrt{-g})_{,\alpha} = \tfrac{1}{2}\sqrt{-g}g^{\mu\nu}g_{\mu\nu,\alpha}. \tag{3.204}$$

With this result we now have

$$\delta(g^{\mu\nu}\sqrt{-g}) = -\left(g^{\mu\alpha}g^{\nu\beta} - \tfrac{1}{2}g^{\mu\nu}g^{\alpha\beta}\right)\sqrt{-g}\,\delta g_{\alpha\beta}. \tag{3.205}$$

Recalling the definition of Einstein's curvature tensor, we have obtained

$$\delta I_g = -\frac{1}{16\pi}\int G^{\alpha\beta}\sqrt{-g}\,\delta g_{\alpha\beta}\,d^4x. \tag{3.206}$$

Thus we derive Einstein's field equation in empty space from the vanishing of the variation of the gravitational action.

If we add the action of matter and radiation fields to the gravitational action, we get the total action. Under arbitrary variations of the gravitational and other fields we insist that the total action vanish. This leads to the Euler–Lagrange equations (3.190) for the matter-radiation fields and to the Einstein equations (3.191) for the gravitational fields. Note, however, that the equations of motion for the matter-radiation fields contain the gravitational fields $g^{\mu\nu}$ and they reduce to the usual form in Minkowski spacetime only when the $g^{\mu\nu}$ can be replaced by the Minkowski tensor.

Return now to the total action (3.189) and vary all fields. Also use the result (3.206) for the variation of $\mathcal{L}_g$

$$\delta I = \int \left[\left\{-\frac{1}{16\pi}G^{\alpha\beta}\sqrt{-g} + \left(\frac{\partial\mathcal{L}_m}{\partial g_{\alpha\beta}} - \partial_\nu\frac{\partial\mathcal{L}_m}{\partial g_{\alpha\beta,\nu}}\right)\right\}\delta g_{\alpha\beta}\right.$$
$$\left. + \left\{\frac{\partial\mathcal{L}_m}{\partial\phi} - \partial_\nu\frac{\partial\mathcal{L}_m}{\partial(\partial_\nu\phi)}\right\}\delta\phi\right]d^4x. \tag{3.207}$$

The last term in each curly brackets was obtained by an integration by parts, as follows:

$$\frac{\partial\mathcal{L}}{\partial f_\mu}\partial f_{,\mu} = \partial_\mu\left(\frac{\partial\mathcal{L}}{\partial f_{,\mu}}\delta f\right) - \partial_\mu\frac{\partial\mathcal{L}}{\partial f_{,\mu}}\delta f,$$

where $f$ stands for either $g_{\mu\nu}$ or $\phi$. The integral over the first term on the right vanishes because the $f$ are not varied on the boundary. Because the variations are otherwise arbitrary, the vanishing of the action implies the vanishing of the coefficients of the varied fields. The variation of the matter fields yields (3.190) (where we have removed $\sqrt{-g}$ because it is unaffected by the $\phi$ variation). The variation of the gravitational fields yields

$$G^{\alpha\beta}\sqrt{-g} = 16\pi\left(\frac{\partial \mathcal{L}_m}{\partial g_{\alpha\beta}} - \partial_\nu\frac{\partial \mathcal{L}_m}{\partial g_{\alpha\beta,\nu}}\right). \tag{3.208}$$

This equation shows how the gravitational and matter fields are coupled. We use it to derive (3.191) by showing that the right side is the energy-momentum tensor in the form (3.192). The familiar form (3.193) then follows. Evaluate the first term on the right side of 3.208:

$$\frac{\partial \mathcal{L}_m}{\partial g_{\alpha\beta}} = L_m\frac{\partial\sqrt{-g}}{\partial g_{\alpha\beta}}. \tag{3.209}$$

Because

$$\frac{\partial g}{\partial g_{\mu\nu}} = gg^{\mu\nu}, \tag{3.210}$$

which follows by differentiating the identity $g = g_{\mu\nu}g^{\mu\nu}$, we obtain

$$\frac{\partial \mathcal{L}_m}{\partial g_{\alpha\beta}} = \left(-\tfrac{1}{2}g^{\alpha\beta}L_m + \frac{\partial L_m}{\partial g_{\alpha\beta}}\right)\sqrt{-g}. \tag{3.211}$$

The matter Lagrangian will not usually depend on derivatives of the metric, but only on the metric itself. Thus, with the above equation we have derived (3.191) with the energy momentum tensor given by (3.192).

## 3.8   Problems for Chapter 3

1. Derive and solve the equations (2.14) for the Lorentz transformation (2.17).

2. Derive the transformation for an arbitrary boost, (2.20).

3. Derive the four-velocity components (2.47).

4. Check that the energy–momentum tensor takes the form (2.54).

5. Derive the coordinate expression for $g_{\mu\nu}$ in (3.4). Review the motion of a free particle in an arbitrary gravitational field, and derive the geodesic equation of motion, (3.38).

6. Derive the transformation property of the Christoffel symbol (3.43).

7. Prove the expression of the affine connection (3.46).

8. Prove the expression involving the Christoffel symbols (3.49).

9. Obtain the geodesic equation (3.60) as the extremal of the propertime.

10. Follow all the steps in the derivation of the $g_{00}$ in (3.72).

11. Derive the expression (3.75) for the covariant derivative of a contravariant vector.

12. Two esoteric-looking results that are used in the variational principle for the derivation of Einstein's equations are (3.101) and (3.102). Derive them in detail.

13. Review the details of the derivation of the conservation of total charge (3.110).

14. Derive (3.113) and with it the expression for the Riemann curvature tensor (3.114).

15. Derive the expression for the Ricci tensor (a contraction of the Riemann tensor) given by (3.122). Show that it is symmetric, though not manifestly so.

16. Prove the Bianchi identities (3.126), thus paving the way to the proof that the Einstein curvature tensor has vanishing covariant divergence (3.128). Prove the latter also. Einstein was unaware of the Bianchi identities and this delayed his discovery of General Relativity.

17. From the above, understand the three possibilities of Section 3.5.

18. Derive at least three of the expressions for the affine connections in static spherical isotropic spacetime (3.141).

19. Derive the Schwarzschild solution for relativistic static stars (3.153).

20. Derive the relationship of the components of the Einstein tensor to the metric functions of static spherical spacetime (3.162).

21. Derive all intermediate steps including the identification of $k$ with the Newton constant $G$ (3.169).

22. Derive the explicit Einstein equations for a static spherical star, (3.172) to (3.175).

23. Go through the details of manipulation of the above equations that are outlined in (3.176) to (3.179).

24. Hence, derive the Oppenheimer–Volkoff equation (3.180).

25. Follow the principle steps in the derivation of the coupled matter and gravitational fields as expressed by (3.190) and (3.191) as given in Section 3.7.1.

# 4

# Compact Stars: From Dwarfs to Black Holes

## 4.1 Birth and Death of Stars

"A star is drawing on some vast reservoir of energy... This reservoir can scarcely be other than the subatomic energy.... There is sufficient in the Sun to maintain its output of heat for 15 billion years."
*A. Eddington, 1920* [46]

Clouds of interstellar gas consisting mostly of molecular hydrogen and a little dust are the incubators of stars. The Horsehead Nebula in Orion is an especially beautiful example. Besides the primordial elements made in the first few minutes in the life of the universe, the clouds contain dust— agglomeration of molecules—that have been spewed from the fiery surfaces of early stars. The clouds are diffuse and highly nonuniform, with clumps and filaments interspersed throughout. The gas spans a wide range of temperature. Most of it is cold at 10 K, but some regions are as hot as 2000 K. Interstellar clouds range in size from less than a light year to several hundred light years across and in mass from 10 to $10^7 M_\odot$. A compression of the order $10^{20}$ is involved in forming a star from this diffuse gas. Stars more massive than a few $M_\odot$ are observed to form in small groups in the densest regions of the clouds. The motion of a given star often suggests the gravitational influence of several nearby stars. About half of all stars are in binaries.

Important factors in stellar formation are gravity, dust, gas pressure, rotation, magnetic fields, winds and radiation from nearby young stars, and radiative shock waves. Our understanding of how these factors conspire to induce collapse of a relatively small region of an otherwise long-lived dynamic, but stable, *molecular cloud* is at a rudimentary level.

The dust in molecular clouds originates on the cool stellar surfaces of *supergiants*, massive stars in a late stage of stellar evolution. Dust shields cloud interiors from ultraviolet starlight, enabling their centers to cool. With lower thermal pressure, gravitational collapse of the denser regions of

TABLE 4.1. Evolutionary times of stars

| $M/M_\odot$ | Years |
|---|---|
| 30 | $5 \times 10^6$ |
| 15 | $1 \times 10^7$ |
| 10 | $2 \times 10^7$ |
| 5 | $7 \times 10^7$ |
| 1 | $1 \times 10^{10}$ |
| 0.1 | $3 \times 10^{12}$ |

clouds becomes inevitable.

At some stage a perturbation (such as the passage of a shock wave or of a shell of expanding gas from a supernova remnant) induces an instability in a critical mass of cloud. Sir James Jeans derived the critical mass (now known as the Jeans mass[1]) in the early twentieth century. A clump of gas then begins a fall toward its mass center under the attraction of gravity. Gravitational energy is converted to heat by the compression. The opacity of the gas increases as its density and establishes a temperature and thermal pressure gradient which approximately balances gravity in a state of quasi-hydrostatic equilibrium. Energy loss by radiation at the *protostar's* surface causes further slow contraction and heating until the core temperature rises to the ignition point for fusing hydrogen into helium ($T \approx 10^7$ K). Thereafter fusion becomes the dominant energy source and the thermal and radiation pressure will now nearly balance gravity for millions to billions of years, a time scale depending approximately on the inverse square of the stellar mass (Table 4.1).

The protostar has now joined the *main sequence* of stars, the evolutionary pathway of stars first proposed by E. Herzsprung and H. N. Russell. The star will spend most of its luminous life in this state of suspended collapse as it burns its large store of hydrogen while slowly radiating energy from its surface.

Thermonuclear fusion[2] provides the energy for the "great furnace," in the phrase of Eddington, which drives stars through the various stages of combustion—hydrogen, helium, carbon, neon, oxygen, magnesium, and silicon. Fusion will end when iron, the most bound nuclear species, is reached. Beyond iron, fusion is no longer exothermic. Nuclei in the region of iron

---

[1]The Jeans mass depends sensitively on the temperature of the gas cloud.

[2]Thermonuclear fusion refers to nuclear fusion induced by quantum mechanical tunneling through the Coulomb barrier between nuclei (first described by G. Gamow). Thermal motion provides the energy for quantum tunneling through a potential barrier that is classically an obstruction. Fusion is exothermic up to iron, ie. the fusion releases heat.

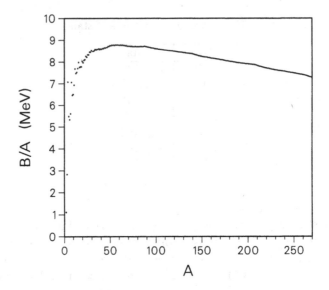

FIGURE 4.1. Binding energy as a function of baryon number showing the peak in binding at iron. (Courtesy of Richard Firestone and the Table of Isotopes Group at Lawrence Berkeley National Laboratory.)

are referred to as the iron peak nuclei because of their higher binding than other nuclei; see Fig. 4.1.

After hydrogen is spent in the core, the next phase, the burning of helium, will commence. Hydrogen in a surrounding shell will continue to burn. The star is hottest in its center and heat is transported by radiation, conduction, and convection to the surface where it is radiated. Ever higher temperatures are required to drive quantum-mechanical nuclear tunneling—first explained by George Gamow—through the increasing Coulomb barriers along the reaction chain. Hydrogen and then helium burning in the core sustain the star on the main sequence for most of its life. As helium burns, a carbon core is formed. Concentric burning shells are established as one element after another is synthesized [47]. Carbon burning in the core lasts for only a few thousand years. Copious gamma rays in the core produce electron–positron pairs that annihilate, producing neutrino pairs. Neutrino losses increase as a high power of the temperature. At this stage the loss is so great that succeeding burning stages progress ever more rapidly—oxygen in a year, silicon in a week. At the exhaustion of each elemental fuel, the core contracts further until the ignition temperature for the next step in the chain is reached. In some cases the heat output from combustion in a succeeding step will be so great as to cause a reexpansion of the outer parts of the star.

The cessation of nuclear fusion signals the end of the luminous stage of

TABLE 4.2. Mass, radius, Schwarzschild radius, and average density of some typical bodies (with all numbers rounded to one significant figure). Nominal neutron star = N.s., white dwarf (Sirius B) = W.d., $M_{Sun} = 1.989 \times 10^{33} g$

| Name | $M/M_\odot$ | R (km) | $r_S$ (km) | $\bar{\rho}$ (gm/cm$^3$) |
|---|---|---|---|---|
| N.s. | 2 | 10 | 6 | $5 \times 10^{14}$ |
| W.d. | 1 | 5400 | 3 | $3 \times 10^6$ |
| Sun | 1 | $7 \times 10^5$ | 3 | 1.4 |
| Jupiter | $10^{-3}$ | $7 \times 10^4$ | $3 \times 10^{-3}$ | 1.3 |
| Earth | $3 \times 10^{-6}$ | 6000 | $9 \times 10^{-6}$ | 5.5 |

the star. The duration of the fusion stage (in which a star spends most of its life) as well as its final evolution depends on the stellar mass. Combustion to the iron end point is attained in massive stars[3] which end through a complex terminal stage either as neutron stars or black holes. Combustion is slower and incomplete in light stars. They end as white dwarfs. Whatever the final end point—dwarf, neutron star, or black hole—it is the gravity of the star's mass that drives stellar evolution and its rate from beginning to end.

For low-mass stars up to a few solar masses, contraction begins again when hydrogen is exhausted in the core. As the temperature increases, hydrogen begins to burn in the outer layers. The envelope expands slowly to become a *red giant* while the core continues to contract and heat. At $10^8$ K three helium nuclei fuse to form carbon followed by an additional helium to form oxygen. These reactions sometimes begin explosively. Eventually, pulsations in the envelope become unstable, and most of the star is shaken off to form a *planetary nebula*.

The remaining core of the star is composed mostly of helium, carbon, oxygen, or magnesium (sometimes a mixture) depending on how far the reactions proceeded before the instability removed the compressing weight of the bulk of the star. With the loss of the envelope, its weight and opacity, the combustion temperature can no longer be sustained. The core now contracts under its own gravity to form a white dwarf[4], a star whose high surface temperature makes it appear white[5]. A white dwarf has a radius of a few thousand kilometers and an average density of the order of $10^6$ that of the Earth (see Table 4.2). After formation, a white dwarf radiates the residue of its earlier hot existence as photons for $10^{10}$ years. It cools,

---

[3]We shall usually use massive stars to refer to those with mass $M \geq 8 M_\odot$.

[4]Most astronomical objects are named after some descriptive feature. The white dwarf was the smallest star known at the time of its discovery.

[5]White dwarfs have surface temperatures as high as 100,000 K and as low as 4000 K, depending on age and the circumstances of birth.

crystallizes, and disappears as a *black dwarf*. It is possible that the universe is not yet old enough for the final state of dwarfs to have formed.

Because of the low mass of the progenitor, gravity propels light stars through their evolution at a much slower rate than massive stars. Stars that terminate their life as dwarfs typically attain great age before the dwarf stage. The Sun, presently of age $4.5 \times 10^9$ years, is expected to live to 12 billion years before collapsing to the dwarf stage, according to modern theory. Eddington's 1920 estimate is in remarkable agreement.

Stars above eight solar masses evolve more rapidly than the progenitors of dwarfs. Most importantly, the thermonuclear reactions proceed further, the star becoming ever hotter until it expands into a super red giant.[6] In the central region of the star, the reactions burn to the iron end point. The core of the star is now supported against collapse only by the pressure of degenerate nonrelativistic electrons. Nuclear burning continues in surrounding shells of Si, O, ... overlying the now inert, central region of iron. The core has a radius of only several thousand kilometers lying within the red supergiant of radius $> 10^8$ km.

Burning in the outer shells adds to the iron core mass. Gravity crushes the core to such a density that electrons become relativistic. The pressure they provide now increases less rapidly with increasing density than was the case at the earlier stage when the electrons were nonrelativistic.[7] Moreover, the kinetic energies of the relativistic electrons have reached the point that capture on protons—*inverse beta decay*—produces an energetically more favorable state. The supporting electron pressure is thus diminished below the point at which it can support further growth in the mass of the iron core against gravity. The core has attained its maximum possible mass named after S. Chandrasekhar who first discovered the limit for an object supported by the pressure of ultrarelativistic, degenerate electrons.

The core then commences a rapid implosion taking less than a second. It becomes extremely hot, attaining temperatures toward the end of collapse of the order of tens of MeV ($\sim 10^{11}$ K). The core is bloated with energetic neutrinos produced by inverse beta decay in the continued *neutronization* of the core material during collapse. The cross-section for the interaction of energetic neutrinos and nuclei at densities of $10^{12}$gm/cm$^3$ is sufficiently large as to trap the neutrinos by collisions in the imploding core. They are swept along with the falling material. Moreover, neutrino pairs are created in great abundance by photoproduction ($2\gamma \longrightarrow \nu\bar{\nu}$) as the collapsing core attains temperatures of tens of MeV. As the core matter is crushed to high density, the Fermi energy of the thermalized electrons and neutrinos rises. Their pressure, together perhaps with the short-range repulsion between nucleons, resists further compression.

The infalling core material rebounds from the stiffened core, sending out-

---

[6]For massive stars the red giant stage is sometimes referred to as supergiant.
[7]See section 4.9.3

ward a shock wave originating somewhere in the core interior. As the shock travels outward, its energy is dissipated by neutrino losses and by photodisintegration of all nuclei in its path. The shock stalls at a few hundred kilometers from the stellar center.

When the stellar material above the region occupied by the core prior to its implosion is no longer supported by the core, a decompression wave travels outward at the speed of sound in the diffuse stellar material of the red giant, and freefall of this material commences. The signal takes a long time to reach the edge of the star compared to the time from the core implosion to shock rebound and its subsequent stall.

The freefalling material is arrested as it meets the stalled shock front, turning the latter into an accretion shock. The infalling matter heats the region of the accretion shock, and at the same time its momentum and mass exert an inward-directed pressure gradient.

A rarefied bubble region develops between the high–density core and the accreting shock front. Neutrino pairs diffusing from the hot interior, annihilate, heat, and expand the bubble. By a complex (and not completely understood) interplay of convection and neutrino heating, a fraction (less than one percent) of the immense gravitational binding energy of the neutron star is transported to the accretion front. It is this small fraction that provides the kinetic energy for the ejection of all but the core of the progenitor star in a *supernova* explosion.

An elementary calculation that we shall shortly do reveals that the energy release of the neutron star (foreseen many years ago by Baade and Zwicky (1934) [1] as the engine driving supernovae) is $\sim 10^{53}$ ergs. It has taken some forty years since the early work of Colgate and White (1966) [48] to arrive at a partial understanding of how a small fraction of this energy release, not much more nor less than 1/100, is so reliably converted to the kinetic energy of ejection (see especially [48, 49, 50]).

For those stellar evolutions that end in a supernova explosion, the hot collapsed core or *protoneutron* star, with temperature of tens of MeV, loses its trapped neutrinos over an interval of some seconds and cools to an MeV or less [51, 50]. At that point the collapsed core has reached its equilibrium composition of neutrons, protons, hyperons, leptons, and possibly quarks. Thus is born a *neutron star* of radius about ten kilometers and average density $10^{14}$ times greater than that of earth. The star continues to cool for millions of years by the slow diffusion of photons to the surface and their radiation into space.

The bulk of the star blown off in the explosion, the *supernova remnant*, expands outward at great velocity (10,000 km/s) sweeping up interstellar gas. The remnant is visible at all wavelengths between radio and X rays, signifying the richness of the processes taking place in the ejecta and between ejecta and the interstellar medium, including molecular clouds. Only in some cases have associations been made between supernova remnants and *pulsars*, rotating magnetized neutron stars [52]–[55]. The remnants are visible for only about $10^4$ to $10^5$ years; pulsars are active for $\sim 10^8$ years or more.

For some unknown fraction of massive collapsing stars, the explosion is believed to fail or to fail to eject sufficient infalling material. The progenitor star therefore continues its *prompt* gravitational collapse to a black hole of mass about equal to the presupernova star ($> 8M_\odot$). This is so because there is a maximum mass for relativistic stars, called the Oppenheimer–Volkoff mass limit, that can be sustained against gravitational collapse by the pressure of degenerate neutrons and their repulsive interaction. Although not many neutron star masses are known, they appear to be clustered around $1.4M_\odot$ [56], close to the Chandrasekhar limit for a degenerate iron core which undergoes implosion at the end of the luminous stage. If the limit for neutron star masses is not far above this, only a small amount of accretion onto the collapsed core will push it beyond the limit.

There is an additional mechanism for *delayed* collapse after formation of a protoneutron star and ejection in a supernova of most of the star. Some of the nucleons and high-energy electrons in the protoneutron star can find a lower energy state as the last of the neutrinos are lost. This phenomenon is *hyperonization*, a process analogous to neutronization, in which the pressure of neutrons and protons is diminished by their conversion to hyperons [57]. Hyperonization may bring about the collapse of neutron stars hovering near the limiting mass [58]–[62].

Thus, continued gravitational collapse can occur under several circumstances. Prompt collapse will result if the explosion fails altogether, creating a *black hole* having the approximate mass of the progenitor, more than $8M_\odot$ and perhaps up to $50M_\odot$. Delayed collapse may occur if the protoneutron star is close to the mass limit even if an explosion does eject most of the progenitor star (see ref. [4]). In the second case, a low-mass black hole ($\sim 1.5$ to $2M_\odot$) will be created. Black holes of about this mass can also be created in the *accretion-induced collapse* of neutron stars in binary systems.[8] Accretion is a slow process. The surface of the neutron star will be heated by infalling material. When the infall rate is such that the radiation pressure from the hot surface counterbalances the gravitational force acting on the infalling material, a limit is reached. It is known as the Eddington limit.

Another type of black hole deserves mention. It is not the end point in the evolution of a single star, but rather is thought to be formed by the condensation of a dense *star-cluster* [63]. Such supermassive black holes as these are probably the maelstroms of *active galactic nuclei*. Supermassive black holes are thought to inhabit the centers of most galaxies. Such a black hole is tearing apart stars as they circle ever nearer the black hole at the

---

[8]Compact stars in binary systems can slowly accrete matter from a less dense companion. Accretion onto a neutron star or white dwarf near their respective mass limits can cause collapse to a black hole or white dwarf, respectively.

center of our own Milky Way before being ingested. This process is signaled by an intense X-ray activity at the galaxy's center.

Once formed, isolated neutron stars and white dwarfs will live almost unchanged forever. They will slowly cool, and their magnetic fields will decay. Pulsar rotation will slow and they will eventually become invisible, their emission terminated when the combination of rotation and magnetic field become too feeble. They will join the host of unseen neutron stars accumulated over the life of the universe. However, the internal baryon constitution of neutron stars will be essentially frozen within a few seconds of birth, once the temperature has fallen below a few MeV.[9] White dwarfs will become crystals after cooling for billions of years, when their temperature falls below the lattice energy associated with nuclei embedded in an electron gas. There is no reason other than an encounter with another star that they should not endure forever as dead stars.

Let us briefly consider another perspective. Of all astrophysical phenomena, none are more essential to life as we know it than the evolution of massive stars and the eventual explosion producing a supernovae. Only a few of the lightest elements were produced during the primordial nucleosynthesis in the Big Bang, none heavier then $^7Li$ [64, 65]. Other elements, up to iron, are synthesized by thermonuclear reactions during the $10^7$ year evolution of massive stars, the heavier of these in the last few days of the life of the presupernova star [47]. The heaviest elements are synthesized after the supernovae explosion in the ejecta which suffers an intense neutrino bombardment from the collapsing core as it releases its binding energy [66].

We ourselves (like the earth and all life on it) are made from the material of earlier generations of stars. Even now, fifteen billion years after the big bang and a thousand or more generations of massive stars latter (table 4.1), the abundance of elements is still dominated by primordial hydrogen and helium, made in the first few minutes in the life of the universe.[10] Nevertheless, the abundance of elements changes ever so slightly with each succeeding stellar generation. The earliest and oldest stars contain negligible amounts of the elements heavier than helium. Younger stars are richer in heavier elements, referred to as their *metalicity*. The lowest energy state of nucleons and nuclei is $^{56}Fe$. This is the ground state toward which the universe is evolving on an unimaginably long timescale. It is interesting to note that since the extinction of the dinosaurs $7 \times 10^7$ years ago, seven generations of massive stars have been born and died.

---

[9]The notable exceptions are millisecond pulsars, which because of the strong centrifugal force, change internal density and therefore composition, as they slowly spin down.

[10]The primordial helium abundance is about 24 percent by mass, coming after hydrogen which is the most abundant element. Heavier elements which are processed in supernovae came much later and appear only in trace amounts.

## 4.2  Aim of this Chapter

In the last chapter we saw that spacetime is curved in our universe, populated as it is by massive objects, galaxies, stars, and so on. This fact, so amazing because of our acquaintance with Euclidean geometry and its apparent accuracy in our vicinity, was arrived at by pursuing the underlying meaning of the equivalence of inertial and gravitational masses. From Schwarzschild's solution to Einstein's equations, we learned the metric of this curved spacetime in the empty space outside a static star.

The interior problem is rather delicate. We are accustomed to thinking of matter in flat spacetime, in terms of the Special Theory of Relativity. Our theories of nuclear matter are worked out in such a frame or even in flat space. By computing properties of matter in flat spacetime (the equation of state, for example), a question of principle must be addressed when the matter resides in strong gravitational fields. Is the local inertial (Lorentz) frame surrounding each spacetime point in the star sufficiently large to justify solving the equations of motion for matter in flat spacetime and, constructing from the solution, the equation of state and the energy-momentum tensor for the relevant densities in the star? If not, we would have to solve the fully coupled problem of gravity simultaneously with the equations of motion containing the general metric functions and the matter and radiation fields (3.190, 3.191).

In this chapter we introduce essential material and concepts for understanding the constitution of compact stars and the various forms in which they may be realized in nature. In the process we answer the question of principle raised in the preceding paragraph. Gravitational units and their relation to the units that are convenient in the nuclear and particle physics of dense matter are introduced. We take the occasion to estimate the mass, baryon number, and radius of neutron stars. Certain general properties of stars and gravity are discussed, including charge neutrality and gravitational redshift.

We introduce elementary models of white dwarfs and neutron stars beginning with the most basic model of dense matter—a free Fermi gas of nucleons. The famous Chandrasekhar asymptotic mass limit of white dwarfs is derived, and then the physical limit imposed by electron capture is discussed. Real white dwarfs are not composed of free nucleons but of nuclei, so carbon and oxygen models are developed.

In the previous chapter we commented on a maximum or limiting mass of compact stars imposed by the nature of gravity. In the present chapter we show that, even if composed of an incompressible medium, a *relativistic* star has a maximum mass.

An upper bound on the limiting mass of neutron stars is obtained from very general considerations and independently of particular theories of dense matter. The limiting mass is important in distinguishing black hole candidates of a few solar masses from neutron stars.

Detailed studies of neutron and other compact stars, such as hyperon stars, hybrid stars with an interior region of quark matter surrounded by nuclear matter, phase transitions, neutron star twins, and black holes are also treated here. The subject of strange stars is developed in *Compact Stars* [4].

## 4.3  Gravitational Units and Neutron Star Size

A compact star near the limit of collapse to a black hole has a higher density than any other star. Such a star is referred to as a neutron star. We want to show that the gravitational field (which in General Relativity is the collection of metric functions $g_{\mu\nu}$) changes by an infinitesimal amount over the distance between nucleons in such a star—and hence for all stars. If this is true, and it is as we shall show, it means that we can solve the problem of the matter fields in flat (Minkowski) spacetime as is customary in treating problems having to do with experiments under conditions such as those on the Earth. This is the physical motivation of the section.

This section also has a practical computational motivation—the introduction of convenient units. We deal in this book with three scales—the astronomical or large scale, the macroscopic or intermediate scale of our own experience, and the microscopic or nuclear and subnuclear scale. For example, the center of a neutron star is composed of matter that is a few times the density of nuclear matter while its surface is composed of iron—a difference of 14 orders of magnitude. The size of a neutron star is of the order of ten kilometers—a measure common to our daily experience. Its total mass, the integral of the energy density over its volume, is up to several solar masses.

### 4.3.1  UNITS

Gravitational units $G = c = k = 1$ (where $k$ is the Boltzmann constant) often are useful, not only because they facilitate computation, but because our equations are not burdened with the appearance of $c$, $G$, and $\hbar$ in various powers and combinations. We start with gravitational (or geometricised) units:

$$1 = c = 2.9979 \times 10^{10} \text{ cm/s} ,$$
$$1 = G = 6.6720 \times 10^{-8} \text{ cm}^3 \text{ g}^{-1} \text{ s}^{-2} , \qquad (4.1)$$
$$1 = k = 1.3807 \times 10^{-16} \text{ erg/K} ,$$

where K denotes temperature in degrees Kelvin. These definitions can be treated as equations so that, for example, we have

$$1 \text{ s} = 2.9979 \times 10^{10} \text{ cm},$$
$$1 \text{ g} = 7.4237 \times 10^{-29} \text{ cm},$$
$$1 \text{ K} = 1.3807 \times 10^{-16} \text{ erg},$$
$$1 \text{ s}^{-2} = 1.4988 \times 10^{7} \text{ g/cm}^{3}, \tag{4.2}$$
$$1 \text{ erg} = 1 \text{ g cm}^2 \text{ s}^{-2} = 8.2601 \times 10^{-50} \text{ cm},$$
$$1 \text{ dyne/cm}^2 = 8.2601 \times 10^{-40} \text{ km}^{-2},$$
$$1 \text{ g/cm}^3 = 7.4237 \times 10^{-19} \text{ km}^{-2}.$$

The last two units are used frequently in astrophysics as units of pressure and energy density, respectively. In particular, note in this last equation that energy density is expressed in inverse squared kilometers. It may seem peculiar that this should be a useful expression. However, if we integrate a quantity with this unit over the volume of a star, it expresses mass in the unit kilometer. As we comment below, the Sun's mass (through the conversion above from grams to centimeters or kilometers) is about 1.5 km. Hence the integral for a neutron star mass will be a number of the order of unity in these units instead of $10^{33}$ g.

In nuclear and particle physics a convenient unit of energy (or mass, $Mc^2$) is a million electron volts or MeV. Sometimes the GeV unit is used because it approximates the nucleon mass of $\sim 939$ MeV. Its value in ergs and other units is given below.

$$\begin{aligned} \text{MeV} &= 1.6022 \times 10^{-6} \text{ erg} = 1.3234 \times 10^{-55} \text{ cm}, \\ &= 1.7827 \times 10^{-27} \text{ g} = 1.1604 \times 10^{10} \text{ K}. \end{aligned} \tag{4.3}$$

The range of the nuclear force is $\sim 10^{-13}$ cm which is defined as a fermi and denoted by fm,

$$1 \text{ fm} = 10^{-13} \text{ cm}. \tag{4.4}$$

Two important constants are the Plank constant $h = 2\pi\hbar$ and the elementary electric charge $e$,

$$\begin{aligned} \hbar c &= 197.33 \text{ MeV fm}, \\ e^2 &= 1.4400 \text{ MeV fm} = (1.3805 \times 10^{-34} \text{ cm})^2. \end{aligned} \tag{4.5}$$

Note the fine-structure constant $e^2/\hbar c$.

From the above value of MeV in grams, we derive energy density as expressed in two popular ways.

$$\begin{aligned} \text{MeV/fm}^3 &= 1.7827 \times 10^{12} \text{ g/cm}^3, \\ &= 1.6022 \times 10^{33} \text{ dyne/cm}^2 \end{aligned} \tag{4.6}$$

The unit on the left is sometimes used in nuclear physics and the one on the right in many astrophysics papers.

In nuclear and particle physics, it is useful to use natural units chosen so that $\hbar = c = 1$. Then from the value of $\hbar c$ written above we obtain by dividing by $fm^4$

$$1/fm^4 = 197.33 \text{ MeV}/fm^3. \tag{4.7}$$

These are both units of energy density and pressure, and both are suitable as stellar quantities, especially at nuclear and supernuclear densities. In the astrophysical literature, energy density is frequently expressed in $g/cm^3$ and pressure in $dyne/cm^2$ ($dyne=erg/cm$). It is therefore useful to have on hand the conversion factors

$$
\begin{aligned}
1/fm^4 &= 3.5178 \times 10^{14} \text{ g}/cm^3, \\
&= 3.1616 \times 10^{35} \text{ dyne}/cm^2. \tag{4.8}
\end{aligned}
$$

Mixed units, which are useful in stellar calculations, are obtained by combining (4.7) with the value of MeV in cm,

$$
\begin{aligned}
1/fm^4 &= 2.6115 \times 10^{-4} /km^2, \\
\text{MeV}/fm^3 &= 1.3234 \times 10^{-6} /km^2. \tag{4.9}
\end{aligned}
$$

The unit on the right side is appropriate for expressing the pressure and energy density when solving the Oppenheimer–Volkoff equations. As noted earlier, when energy density of a neutron star, expressed in such units is integrated over the radial coordinate, it yields a mass in the unit of kilometers. This can be compared with the solar mass,

$$M_\odot = 1.4766 \text{ km} = 1.989 \times 10^{33} \text{ g} = 1.116 \times 10^{60} \text{ MeV}. \tag{4.10}$$

The binding energy of the Sun is small compared to its mass because it is diffuse. Therefore we can estimate the number of nucleons in the Sun as

$$N_\odot = \frac{M_\odot}{m_n} \approx 1.1 \times \frac{10^{60}}{940} \approx 10^{57} \tag{4.11}$$

The number of nucleons in a neutron star would be of the same order.

Let us compare the spin of a neutron star expressed in units of $\hbar$, for curiosity. We note first that

$$\hbar = 2.6115 \times 10^{-76} \text{ km}^2. \tag{4.12}$$

From the latter we can calculate the spin of a pulsar in units of $\hbar$. In reference [4], we find for the moment of inertia of the Crab pulsar, $I \sim 70 \text{ km}^3$. Its period of rotation is $P = 1/30 \text{ s} = (2\pi)/\Omega$. We find it has a spin of

$$
\begin{aligned}
J = I\Omega &= 1.3 \times 10^4 \text{ km}^3/s = 1.3 \times 10^4 \text{ km}^2 \cdot \text{km/s} \\
&= 5 \times 10^{79}\hbar. \tag{4.13}
\end{aligned}
$$

This has no practical use, but is an interesting contrast with the spins of nuclei that range up to several tens of $\hbar$ for very excited states.

At various places in the book, we shall have occasion to use some of the relations derived above. Dimensional analysis will always aid in converting a computed quantity in gravitational or natural units back to familiar units that express its dimensions. For example, if in a computation it is found that a particle has an acceleration of $a = x/\text{km}$, knowing the dimension of $a$ as length/time$^2$, we can multiply $a$ by whatever quantities have been set to unity, so as to regain the dimension of $a$. Thus,

$$a = x/\text{km} = xc^2/\text{km} = (x/\text{km})(3 \times 10^5 \text{ km/s})^2 = 9x10^{10} \text{ km/s}^2 . \qquad (4.14)$$

Of course sometimes one needs to use several of the quantities that have been set to unity, with various powers. At other times, relations between units derived above can be used. We encounter examples below, where now we return to an estimate of the size and number of baryons contained in a star.

### 4.3.2  SIZE AND NUMBER OF BARYONS IN A STAR

We have noted that the Schwarzschild metric becomes singular at

$$r = r_S \equiv 2M \qquad (\text{upon setting G} = c = 1) . \qquad (4.15)$$

For actual stars this radius is interior to the star itself where the Schwarzschild solution does not hold and the singularity therefore does not exist. However, for a neutron star, the Schwarzschild radius is not deep within the star as compared to other heavenly bodies as shown in Table 4.2. But in the special case where the star lies within its "gravitational radius," it must be a black hole.

Next we estimate the mass and radius of a star near the limit of gravitational collapse having $R = 2M$. This will be an estimate of the limiting or maximum mass possible for a neutron star. Assume that gravity packs nucleons up to their repulsive cores, say $r_0 \approx 0.5 \times 10^{-13}$ cm. Then

$$R \approx r_0 A^{1/3}, \quad M \approx Am , \qquad (4.16)$$

where $A$ is the number of baryons in the star and $m$ is their mass,

$$m \approx 939 \text{ MeV} = 1.7 \times 10^{-24} \text{ g} = 1.2 \times 10^{-52} \text{ cm} . \qquad (4.17)$$

Hence, substituting (4.16) into the equation $R = 2M$, we find

$$A^{2/3} = r_0/(2m) = 1.9 \times 10^{38} . \qquad (4.18)$$

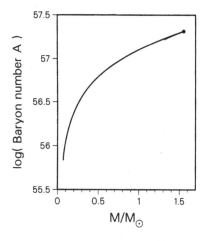

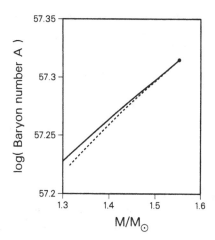

FIGURE 4.2. Total number of baryons vs. mass for neutron stars. See Chapter [4] for details. ($K$=240 MeV, $m^*_{\text{sat.}} = 0.78$, $x_\sigma = 0.6$).

FIGURE 4.3. Enlargement of Fig. 4.2 in the vicinity of the limiting star. Dotted line corresponds to unstable stars above the limiting mass star in density.

Putting this result back into the expressions for radius and mass, we have

$$
\begin{aligned}
A &= 2.6 \times 10^{57}, \\
R &= r_0 A^{1/3} = 7 \text{ km}, \\
M &= R/2 = 3.5 \text{ km} = 2.3 M_\odot.
\end{aligned}
\tag{4.19}
$$

Here we have an estimate of the baryon number, radius, and mass of a star at the limit of collapse to a black hole. Consequently, as regards a typical neutron star, its radius will be larger and its mass and baryon number smaller than these numbers. In Figs. 4.2 and 4.3 the baryon number is plotted as a function of stellar mass for a model sequence of neutron stars. Later in this chapter we discuss how this model sequence can be calculated.

Because the matter at the edge of the star will be less compacted than at the center, we expect a somewhat larger radius and smaller mass, say

$$
M \approx 2 M_\odot, \quad R \approx 10 \text{ km}.
\tag{4.20}
$$

The average density of such an object is

$$
\bar{\epsilon} = 10^{15} \text{ g/cm}^3.
\tag{4.21}
$$

It is interesting to compare this result with the density of normal symmetric nuclear matter. Taking the average of

$$
m_n = 939.57 \text{ MeV}, \quad m_p = 938.28 \text{ MeV},
\tag{4.22}
$$

and using the saturation baryon density and binding per nucleon of nuclear matter obtained from the large $A$-limit of the droplet model of nuclear masses [67],

$$\rho_0 = 0.153 \text{ baryons/fm}^3, \qquad -B/A = 16.3 \text{ MeV}, \qquad (4.23)$$

we get for the energy density of normal symmetric nuclear matter

$$\epsilon_0 = 141 \text{ MeV/fm}^3 = 2.51 \times 10^{14} \text{ g/cm}^3. \qquad (4.24)$$

Compare this with the average density of earth which is 5.5 grams per cubic centemeter.

The above crude estimate of the gross properties of a compact star near the limit of gravitational collapse suggests that the average density is about four times nuclear matter density. The density of the star near the mass limit is supernuclear. Also a star must be charge neutral because the repulsive Coulomb force for an infinitesimal charge per baryon in the star will expel particles of like charge. Therefore neutrons will be the majority population—hence, the name "neutron" star.

### 4.3.3   GRAVITATIONAL ENERGY OF A NEUTRON STAR

We can roughly estimate the gravitational energy for a uniform density neutron star with a nominal two solar mass and a 10-km radius such as discussed above. Using (3.167) we find

$$M(r) = \frac{4\pi}{3}r^3\bar{\epsilon}, \quad M = M(R),$$

$$E_G \approx \int_0^R \frac{M(r)\, dM(r)}{r} = \int_0^R \frac{M(r)(4\pi r^2 dr\bar{\epsilon})}{r} = \frac{3}{5}\frac{M^2}{R}$$

$$= 0.54 \text{ km} = 6.5 \times 10^{53} \text{ erg} = 7.3 \times 10^{32} \text{ g}$$

$$= 4.1 \times 10^{59} \text{ MeV}. \qquad (4.25)$$

Thus, the gravitational energy is a significant fraction of the total gravitational mass of the star, namely, $0.54/3 = 0.18$, or 18% for a two solar mass star ($M_\odot \approx 1.5$ km). To find the gravitational energy per nucleon in the star, refer above to the estimated number of nucleons $A$ to find 157 MeV per nucleon. By contrast, the binding of a nucleon in nuclear matter is about 16 MeV and in a finite nucleus it is about 8 MeV.

The classical estimate of the gravitational energy is not the binding energy of the star; it does not take account of the compression energy resisting the assembly of nucleons in a star arising from the Fermi pressure and the short-range repulsive nuclear force. Nonetheless, the binding energy per nucleon in a limiting mass star is about 100 MeV per nucleon (see Fig. 4.16). Thus, the weakest force binds nucleons in a neutron star ten times more strongly than the strong force binds them in nuclei. How can this be so?

The answer lies in their ranges. The nuclear force acts only on neighbors that lie within its range, typically the several next nearest neighbors, but each particle in the star feels the gravitational force of all others.

From the absolute limit on $M/R$ of a stable star derived in Section 4.16, we can find an upper limit on the gravitational energy of a star relative to its mass. It is $(3/5)(M/R) < 4/15$ or 27%. The import of the large binding of a neutron star is that the energy released during a few seconds in a supernova event is large compared to that produced in thermonuclear fusion reactions during the entire stellar lifetime of a few million years (Table 4.1).

## 4.4   Partial Decoupling of Matter from Gravity

In nuclear and particle physics, we do not usually think of the matter Lagrangian and matter field equations as depending on the metric because we usually ignore gravity. In this event the metric tensor is the diagonal, constant, Minkowski metric $\eta_{\mu\nu}$ (2.9). For example, the Dirac equation for a nucleon contains the scalar product $\gamma_\mu \partial^\mu$, and the Lagrangian for a scalar meson contains $\partial_\mu \phi \partial^\mu \phi$. But these are merely shorthand notations that avoid the necessity of writing the Minkowski metric explicitly in the scalar product of four-vectors, $\gamma_\mu \partial^\mu = \eta_{\mu\nu} \gamma^\nu \partial^\mu$. They are correct when the metric $g_{\mu\nu}$ can be justifiably replaced by the Minkowski metric of a Lorentz frame. However, this can be done only in a sufficiently small locality, the smallness of which depends on the curvature of spacetime. It is not possible to erect a global Lorentz frame. The differential equations for the gravitational fields, of which the $g_{\mu\nu}(x^\alpha)$ are the generalizations of the Newtonian gravitational potential, and those of the matter and radiation fields are therefore coupled in general [(3.190) (3.191)].

For applications in strong gravitational fields we need to determine whether it is justified to solve the matter equations of nuclear and particle physics in flat spacetime. For stellar structure, this amounts to ascertaining the relevance of an equation of state for neutron star structure (and hence for all stars), one that is derived in a Lorentz frame. Let us put the mass and radius of a star near the limit of gravitational collapse into the metric for the space outside a star (3.153). We use the fact that, at the origin, $g_{11} = -1$ [see (3.171)] to obtain

$$\frac{g_{11}(R)}{g_{11}(0)} = \left(1 - \frac{2M}{R}\right)^{-1} = \left(1 - \frac{6}{10}\right)^{-1} = \frac{10}{4}. \tag{4.26}$$

The metric changes by a factor 2.5 over the dimension of the star. Therefore the relative change in the metric over the spacing of nucleons in the star is $\sim 2r_0/R = 2A^{-1/3} \approx 10^{-19}$. This is a negligible change over a distance spanning many times the internucleon spacing. From this it is clear that a volume over which conditions do not change significantly can be defined for every region of the star. The pressure is certainly different in regions that differ in radial location by an appreciable amount. But the conditions for defining the local state of affairs are fully met in all such volumes.

However, such local regions are not inertial regions. The second two of the conditions for local Lorentz frames defined in Section 3.2.3 are satisfied, but not the first. Rather, the gravitational field is finite though uniform in each such region. Therefore we must consider further the question of validity of the notion of an equation of state of matter that describes the role of nuclear, atomic, and molecular physics in the context of the interior of a star. This was done first in the 1965 pioneering work of Wheeler and his collaborators on neutron stars [43]. We follow their reasoning.

Imagine an idealized hollow tube extending from the center of the star through one such region as above. Let a small laboratory be propelled upward in the tube at such a velocity that, moving freely, it comes to rest before beginning its descent at just the location of the region in question. From this locally inertial laboratory the volume of the adjacent region and the number of baryons it contains can be measured and therefore also its baryon density. We are interested in stars composed of matter in its lowest energy state and that is charge neutral and at zero temperature.[11] Therefore, all equivalent regions having the same number of baryons and volume also have the same energy, or energy density. The energy density in all small regions of the star is therefore uniquely specified by the baryon density,

$$\epsilon = \epsilon(\rho) , \tag{4.27}$$

where $\epsilon$ denotes energy density and $\rho$ baryon density. If we enlarge the tube spoken of above, so that one of the regions can fall freely, it occupies an inertial frame, and the uniqueness in the definition of cold catalyzed matter assures that the properties of matter can be computed in a Lorentz frame to arrive at the equation of state, (4.27).

Thus, each small volume in the star, which, however, is large enough to be uniform in density, can be described by the laws of physics of the Special Theory of Relativity to high accuracy. We make a negligible error for neutron stars by solving the field equations for matter in the absence of gravity (i.e., in flat spacetime). From the solution in Minkowski spacetime, we can construct the energy-momentum tensor that is diagonal in a comoving Lorentz frame (2.52):

$$T^{\mu\nu} = \begin{pmatrix} \epsilon & 0 & 0 & 0 \\ 0 & p & 0 & 0 \\ 0 & 0 & p & 0 \\ 0 & 0 & 0 & p \end{pmatrix} . \tag{4.28}$$

The principle of general covariance (Section 3.6.4) tells us how to construct the energy-momentum tensor in an arbitrary gravitational field from

---

[11]Such matter, we call *neutron star matter* or *cold catalyzed matter*, matter from which all energy that can be extracted has been extracted.

the one derived in Minkowski spacetime. Of course, the general energy-momentum tensor (3.156) contains the metric tensor, which is the gravitational field (but not yet determined). We find how the matter of our theory is arranged under the influence of gravity and how in turn it shapes the gravitational fields $g_{\mu\nu}(x)$ by putting a solution to the matter problem found in Minkowski spacetime into the general energy-momentum tensor on the right side of Einstein's field equations (3.132). The Oppenheimer–Volkoff equations are the result of just such an operation.

## 4.5   Equations of Relativistic Stellar Structure

### 4.5.1   INTERPRETATION

In Chapter 3 we found that Newton's equations of stellar structure are,

$$4\pi r^2 dp(r) \;=\; -\frac{M(r)dM(r)}{r^2} \tag{4.29}$$

$$dM(r) \;=\; 4\pi r^2 \epsilon(r)\, dr \tag{4.30}$$

By comparison, in Chapter 3 we saw that for a static isotropic star, Einstein's equations take a special form (3.167, 3.180), derived first by Oppenheimer and Volkoff. By rewriting them we can see the Newtonian part and arrive at the physical interpretation:

$$4\pi r^2 dp(r) \;=\; -\frac{M(r)dM(r)}{r^2} \tag{4.31}$$
$$\left(1+\frac{p(r)}{\epsilon(r)}\right)\left(1+\frac{4\pi r^3 p(r)}{M(r)}\right)\left(1-\frac{2M(r)}{r}\right)^{-1},$$

$$dM(r) \;=\; 4\pi r^2 \epsilon(r)\, dr. \tag{4.32}$$

The interpretation is very simple. The first of the pair of equations expresses the balance between the force acting on a shell of matter due to material pressure from within and the weight of matter weighing down on it from without. Think of a shell of matter in the star of radius $r$ and thickness $dr$. The second equation gives the mass–energy in this shell. The pressure of matter at the inner edge of the shell is $p(r)$ and at the outer edge, $p(r)+dp(r)$. The left side of the first equation is the net force acting outward on the surface of the shell by the pressure difference $dp(r)$, and the first factor on the right side is the attractive Newtonian force of gravity acting on the shell by the mass interior to it. The remaining three factors are the exact corrections for General Relativity. So these equations express the balance at each $r$ between the internal pressure as it supports the overlaying material against the gravitational attraction of the mass-energy interior to $r$. They are the equations of hydrostatic equilibrium in General Relativity.

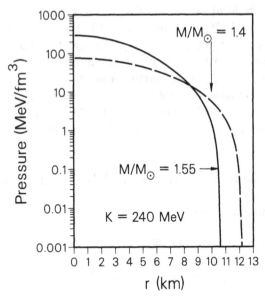

FIGURE 4.4. Illustration of the pressure distribution in two stars of neighboring mass, showing the steeper pressure gradient with increased mass and the corresponding reduction in radius thus, illustrating a mass limit above which the radius will be driven below the Schwarzschild limit. The actual limit for stable stars occurs before the Schwarzschild radius is reached.

Why did the equations for stellar structure lead us to the equation for hydrostatic equilibrium? Because they were built in by the condition that each fluid element is at rest in the star (3.163).

The differential equations (4.31,4.32) assures that the pressure decreases monotonically in a star. This is clear because the derivative of pressure is negative. It also makes sense because the amount of overlaying material decreases with radial coordinate, meaning that the greatest weight weighs on the center and decreases in the outward direction. The equation of state $p = p(\epsilon)$ is the manner in which the properties of dense matter enter the equations of stellar structure. Otherwise they are completely specified. It is a property of matter that the pressure never decreases as a function of increasing density. The pressure can remain constant over an interval in density as in first-order phase transitions of simple (one-component) substance, but otherwise increases with density [68]. Therefore we may conclude that the density of matter in a star is also a decreasing function of the radial coordinate.

It is a particularly notable feature of the Oppenheimer–Volkoff equations that the internal pressure of matter appears together with the energy density on the right side of equation (4.31), thus sharpening the pressure gradient in a star, compressing matter, and reducing the stellar radius along the sequence of stars with increasing central density. The radius is ultimately

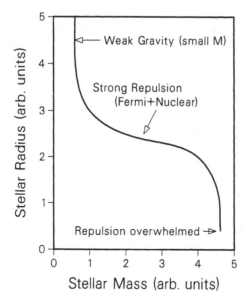

FIGURE 4.5. Mass–radius relationship that is characteristic of degenerate stars—large radius for low mass, smallest radius for star at the mass limit.

reduced to a value below the Schwarzschild limit (see Fig. 4.4). As a consequence, no matter how large the pressure may be, it ultimately cannot prevent the collapse of stars whose mass exceeds what is referred to as the *limiting mass* (Section 3.6.6).

Closely related to the above discussion is the characteristic relationship between the mass of a degenerate relativistic star and its radius. The smaller the mass, the less the gravitational attraction and the larger the radius. At intermediate mass, the Fermi pressure of electrons in the case of white dwarfs, and of nucleons and their mutual repulsion when in close proximity as in neutron stars, resists gravity, leading to a range of masses for which the radius hardly changes. Finally, as the limiting mass is approached, the radius decreases sharply with mass (Fig. 4.5).

Newtonian stellar structure is obtained as a special case of the Oppenheimer–Volkoff equations by dropping the last three factors in (4.31). In Newtonian stars, $\epsilon$ is dominated by the rest mass of the baryons which, however, does not contribute to the pressure. Therefore, the first and second terms of the second line of (4.31) are close to unity for the same reason, namely, that $p \ll \epsilon$. As noted in the first chapter, the last term is always positive for stars. Indeed for stars such as our Sun, $2M/R = 4.3 \times 10^{-6}$, and so the last factor is also unity for Newtonian stars.

The correctness of Einstein's theory, and therefore the equations of stellar structure, is confirmed by the observational tests of General Relativity. The most impressive of these are concerned with the Hulse–Taylor pulsar binary

[69], known by its celestial coordinates as 1913+16 and studied with great care since its discovery in 1974 [19, 70, 71]. The orbit's orientation precesses in the strong gravitational field of the neutron stars at a rate of more than 4 degrees per year. This compares with the former classic confirmation of General Relativity—the general relativistic contribution to the precession of Mercury—which is only 43 seconds per century.

Einstein's gravitational radiation prediction, which affects the orbital motion of the binary pulsar pair causing the orbit to decay, is confirmed to less than one percent [19]. All other variants of Einstein's theory of gravitation are ruled out at this level of accuracy. At the measured rate of decay of the orbit due to the radiation of gravity waves, the two members of the binary will collide in about $10^8$ years. While this is a long time, it is to be compared with the galaxy age of $10^{10}$ years. Collisions of compact stars are therefore likely events in the universe, and are candidates for observation in future generations of gravitational wave detectors. Such collisions are also of possible importance to the question of strange quark nuggets as cosmic radiation [4].

### 4.5.2   BOUNDARY CONDITIONS AND STELLAR SEQUENCES

The two Oppenheimer–Volkoff equations can be integrated from the origin $r = 0$ with the initial conditions $M(0) = 0$ [since near $r = 0$ we may write $M(r) = (4/3)\pi r^3 \epsilon(0)$] and an arbitrary value for the central energy density $\epsilon(0) = \epsilon_c$. The equations are to be integrated until the pressure $p(r)$ becomes zero. Zero pressure can support no overlying material against the gravitational attraction exerted on it from the mass within and so marks the edge of the star. The point $R$ where the pressure vanishes defines the radius of the star and $M(R)$ its gravitational mass.

For a given equation of state, there is a unique relationship between the stellar mass and central density $\epsilon(0)$ because the Oppenheimer–Volkoff equation is of first order. Thus, for each possible equation of state, there is a unique *sequence* of stars parameterized by the central density or, equivalently, the central pressure. For a vast range of densities spanning many orders of magnitude, a sequence of stars from white dwarfs to neutron stars is shown in Fig. 4.6.

The white dwarfs are the lowest density sequence of compact stable stars. Their central densities range from about $\sim 5 \times 10^5$ to $9 \times 10^9$ gm/cm$^3$. The more massive ones have densities that are $\sim 2 \times 10^9$ times greater than the average density of Earth. White dwarfs consist of isotopes from hydrogen to carbon and oxygen with possible admixtures of some heavier ones. All atoms, save at the surface, are ionized by the high compression. As the mass rises above $\sim M_\odot$, depending on the particular equation of state used to describe the matter of white dwarfs, they become unstable, and the sequence ends. The conditions for stability will be studied in Section 4.19. But we may note here, in connection with Fig. 4.6, that the solid parts of

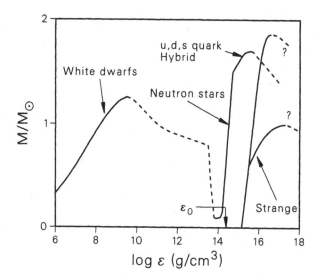

FIGURE 4.6. Schematic illustration of the solution of the O–V equations over a broad range of central densities. Hypothetical strange stars are also shown. (cf. [4]). Solid lines—where the slope is positive—are stable configurations. See discussion of stability in the text (Section 4.19).

the curves—those with $dM/d\epsilon > 0$—represent stable configurations.

Stable white dwarfs are supported against gravitational collapse by the pressure of degenerate electrons. Degeneracy refers to a situation for which the temperature is much smaller than the Fermi energy of the electrons, so that electrons essentially fill all momentum states allowed by the Pauli principle up to a maximum value, sometimes referred to as the Fermi momentum. However, with increasing white dwarf mass (and therefore increasing central density and pressure), the electron Fermi energy rises. A point is reached at which it becomes energetically more favorable for electrons to be captured by protons to form neutrons by inverse beta decay. The supporting pressure of the relativistic electrons is lost, and a region of a few decades in central energy density follows in which no stable stars exist between white dwarfs and neutron stars, as illustrated in Fig. 4.6.

The Fermi pressure of degenerate neutrons reestablishes stability against collapse for the lightest neutron stars. For more massive stars with higher densities, the repulsion of the nuclear force at short range adds additional resistance to compression.

Neutron stars are denser by far than white dwarfs. Neutron stars are five to ten times more dense at their centers than the nuclei of atoms ($\rho_0 \sim 2.5 \times 10^{14}$ gm/cm$^3$). They are about $3 \times 10^{14}$ more dense than Earth. The density falls from its high central value by some fifteen orders of magnitude to the surface.

It should be clear (but well worth emphasizing) that neutron stars are held together by gravitational attraction, not by the nuclear force. After all, nuclei themselves are stable only for $A \lesssim 250$ and neutron matter not at all. Indeed, except possibly for the very lightest neutron stars, gravity has compacted the neutron matter so severely that the most important part of the nuclear force is the repulsion. If gravity were switched off, a neutron star would explode! As well as white dwarfs and neutron stars, hypothetical strange star sequences are shown. Being hypothetical, their mass limit is unknown.

As remarked above, the edge of a star occurs at the point where $p(R) = 0$. Frequently, one assumes that the pressure goes to zero as the energy density. But in cold matter, nucleons that are condensed into nuclei are of lower energy than a gas of nucleons. Material made of nuclei, like iron or carbon, has zero pressure because it is in equilibrium. So, the edge of an ideal *cold* neutron star or white dwarf occurs at a finite energy density corresponding more or less to the laboratory density of whatever nuclei occur in the edge of the star. As we will discuss later, the prevalent species will depend on the progenitor star in the case of a white dwarf. The dominant species will be iron atoms in a Coulomb lattice (density $\sim 7.85$ gm/cm$^3$ ) in the case of a neutron star. We shall discuss the low density range in Section 4.10 in connection with the equation of state of subnuclear matter.

The effects of finite temperature, which in a typical neutron star is far higher than can be achieved in the laboratory, will be discussed with reference to the nature of the surface material in Section 4.11. However, the surface region contributes so little to the large-scale properties of a compact star that the equation of state of cold matter in the ground state is generally employed.

## 4.6    Electrical Neutrality of Stars

The repulsive Coulomb force acting on any charged particle of the same sign as the net charge on a star will overwhelm the gravitational attraction acting on the additional particle, and it will be expelled unless the net charge on the star lies below a certain minimum value which we wish to calculate. The condition that an additional like charge will not be expelled is given by

$$\frac{(Z_{\text{net}} e)e}{R^2} \leq \frac{GMm}{R^2} < \frac{G(Am)m}{R^2},\qquad(4.33)$$

where $Z_{\text{net}}$, $R$, $M$, are the net charge on the star, its radius, and mass, respectively, while $m$ and $e$ the mass of a proton and its charge. The number of baryons in the star is denoted by $A$, so that we have $M < Am$, that is, the mass of the star, because of its gravitational binding, is less than the mass of the baryons distributed to infinity. Hence (in gravitational units),

$$Z_{\text{net}}/A < (m/e)^2 \,. \tag{4.34}$$

A net charge larger than the above value would not allow any additional charged particle of the same sign to be gravitationally bound.

We evaluate $(m/e)^2$ for a proton. In that case using (4.5),

$$\left(\frac{m}{e}\right)^2 \sim \frac{(938 \text{ MeV})^2}{1.44 \text{ MeV fm}} \sim 10^{-36} \,. \tag{4.35}$$

Hence,

$$Z_{\text{net}} < 10^{-36} A \,. \tag{4.36}$$

This is the limit on the net positive charge. If the star had a net negative charge, we could similarly compute its limit considering the balance of Coulomb and gravitational forces on an added electron. The limit would be reduced by the factor $m_e/m$.

We conclude that the net charge per nucleon (and therefore the average charge per nucleon on any star) must be very small, essentially zero. However, the number of charges of equal and opposite sign is not at all limited. It will be important in another context to note that the above condition on net charge on a star is a global one and not a local one. The above result places no restriction on the value of the charge density as a function of location in the star, so long as it integrates to less than the small value derived above.

## 4.7  "Constancy" of the Chemical Potential

Let us recall several thermodynamic relationships. The pressure is defined as

$$p = -\frac{\partial E}{\partial V} \,, \tag{4.37}$$

taken at constant entropy, where $E$ is the energy and $V$ the volume. For the present, we consider stars that are cold on the nuclear scale. The energy density and baryon number density are

$$\epsilon = E/V, $$
$$\rho = A/V \,, \tag{4.38}$$

where $A$ is the number of conserved baryons in the volume $V$, which might be a locally inertial region. Suppose the equation of state is known in the form

$$p = p(\rho), \qquad \epsilon = \epsilon(\rho) \,, \tag{4.39}$$

from which $\rho$ can always be eliminated, if desired, to provide $p = p(\epsilon)$. Now we can write

$$p = -\frac{\partial(\epsilon/\rho)}{\partial(1/\rho)} = \rho^2 \frac{\partial}{\partial\rho}\left(\frac{\epsilon}{\rho}\right) = \rho\mu - \epsilon, \tag{4.40}$$

where

$$\mu \equiv \frac{d\epsilon}{d\rho} \tag{4.41}$$

is the baryon chemical potential. From the relationship above between $p$ and $\mu$, we find

$$\frac{dp}{d\rho} = \rho\frac{d\mu}{d\rho}. \tag{4.42}$$

The Oppenheimer–Volkoff equation in the form (3.185) can be written

$$-\int_r^{r'} d\nu = \int_r^{r'} \frac{dp}{p+\epsilon} = \int_r^{r'} \frac{d\mu}{\mu}, \tag{4.43}$$

where $r < r' < R$, with $R$ the radius of the star. Because $r$ and $r'$ are otherwise arbitrary, we obtain

$$\mu(r)e^{\nu(r)} = \mu(r')e^{\nu(r')} = \text{const} \tag{4.44}$$

This shows that the baryon chemical potential at any depth in a star, corrected by the factor $e^{\nu(r)}$, is a constant. (A theorem for an arbitrary number of chemical potentials of independent components has been proven by Kodama [72].) The correction factor for energy is the inverse of the gravitational time dilation factor we will encounter in the next section.

The pressure $p$ necessarily vanishes at the stellar surface. Let us think what the surface material of a "neutron" star made from the processed material of an evolved star could be. Surely the lowest energy equilibrium state ($p = 0$) of hadronic matter is $Fe^{56}$ which is the end point of fusion reactions in presupernova stars. Solid iron in its equilibrium state has $p = 0$. Iron therefore forms the surface of an idealized *cold* neutron star. From (4.38) and (4.40) we find that

$$\frac{E}{A} = \left(\frac{\epsilon}{\rho}\right)_0 = \mu_{Fe}. \tag{4.45}$$

Hence, we obtain

$$\mu_{Fe} \approx \frac{1}{56}m(Fe^{56}) = 930.54 \text{ MeV}, \tag{4.46}$$

where $m(Fe^{56})$ is the mass of the iron atom and we have ignored small energies associated with the molecular forces of the solid. Consequently, anywhere in the star it must be true that

$$\mu(r)e^{\nu(r)} = \mu_{Fe}\left(1 - \frac{2M}{R}\right)^{1/2}. \tag{4.47}$$

This is not a condition that has to be imposed on the solution of the Oppenheimer–Volkoff equation. Rather it is the way that matter is arranged by gravity. From (4.44) we learn that, in the equilibrium configuration of the star, a particle can be brought from infinity to any point in the star, with the same gain in gravitational energy. In fact, (4.44) holds not only for the chemical potential, but also for the temperature at various points in a star in equilibrium. Thermal equilibrium does not correspond to $T =$ const but to $T(r)e^{\nu(r)} =$ const (cf. [73], p. 263).

## 4.8 Gravitational Redshift

The dilation in time, measured by observers in relative uniform motion, is familiar from the Special Theory of Relativity. Such an effect of a different origin holds also for stationary observers in a static gravitational field located at positions of different gravitational potential.

### 4.8.1 INTEGRITY OF AN ATOM IN STRONG FIELDS

Consider the question of gravitational effects in the context of comparing spectral lines of atoms located on distant stars with those of atoms located on Earth. In Chapter 3 we saw that, after a photon is emitted, its frequency will be shifted if it moves from one place to another having different gravitational fields. But is the internal structure of the atom itself unaffected by the gravitational field in which it is located, especially if the field is very strong?

Let us compare the change in the acceleration due to gravity across the diameter of the atom with the acceleration of the electron in the atom by the Coulomb force. The (Newtonian) acceleration due to gravity at the surface of a star is

$$g_\star = \frac{M}{R^2} \leq \frac{1}{2}\frac{1}{R}, \tag{4.48}$$

where we have replaced $M/R$ by $1/2$ to obtain an upper limit. Again, taking $R = 10^6$ cm and $1 = c = 3 \times 10^{10}$ cm/s, we have

$$g_\star < \frac{1}{2}\frac{1}{10^6\ \text{cm}}\left(3 \times 10^{10}\ \text{cm/s}\right)^2 = 4.5 \times 10^{14}\ \text{cm/s}^2. \tag{4.49}$$

(Notice how we have converted back from units with $c = 1$ in which acceleration is expressed in units of inverse length or, equivalently, inverse time to ordinary units of cm/s$^2$.) The difference of the acceleration across the dimension of the atom is

$$\delta g_\star = \frac{M}{(R-r)^2} - \frac{M}{(R+r)^2} \approx \frac{4r}{R}g_\star. \tag{4.50}$$

We obtain

$$\delta g_\star \approx 2 \times 10^{-14} g_\star = 9 \text{ cm/s}^2 . \tag{4.51}$$

By comparison, the acceleration of the electron in a helium atom in a Bohr orbit is

$$a = \frac{1}{m_e} \frac{e^2}{r_{\text{Bohr}}^2} \approx \frac{1.44 \text{ MeV fm}}{0.5 \text{ MeV}(\frac{1}{2}10^{-8} \text{ cm})^2} \approx 10^{25} \text{ cm/s}^2 . \tag{4.52}$$

So the acceleration experienced by the electron in the hydrogen atom is $\sim 10^{24}$ times larger than the change in the gravitational acceleration across the orbit when the atom is located on the surface of the star. It is also much larger than the gravitational acceleration at the surface of the star. We reach the conclusion that even the strongest gravitational fields on stars cannot affect the workings of an atom (much less a nucleus).

## 4.8.2    REDSHIFT IN A GENERAL STATIC FIELD

Now consider an idealized situation in which we have an observer both at a great distance from a static star and from every other body and at rest with respect to the star. The gravitational field is therefore static. Let the mass and radius of the star be $M$ and $R$, respectively. Let two identical atoms be located, one at the surface of the star and the other with the distant observer whose location we shall denote by $\infty$. For later simplicity, but by no means essential, let the center of the star, the atom on its surface, and the distant observer lie along the same line. Think of the emission of a photon as corresponding to a wave train. The two ends of a cycle, or equivalently, successive wave crests emanating from the source, correspond to neighboring spacetime events.

Choose a local inertial frame in which the atom is at rest. Then,

$$d\tau_{\text{a}} = \sqrt{\eta_{00}} \, d\xi_a^0 = d\xi_a^0 \tag{4.53}$$

is the invariant interval of emission from the atom (subscript "$a$") and is equal to the coordinate time $d\xi_a^0$ between wave crests in this local inertial frame. This relationship holds in the strong field of the star as at the location $\infty$ of the observer, as proven in the above section. In an arbitrary gravitational field, the spacetime interval $dx^\mu$ between crests is governed by

$$d\tau_{\text{a}} = (g_{\mu\nu} dx^\mu dx^\nu)^{1/2} . \tag{4.54}$$

In particular, for the atom at rest on the surface of the star,

$$d\tau_{\text{a}} = \sqrt{g_{00}(R)} \, dx^0 \tag{4.55}$$

gives the coordinate time interval $dx^0$ between crests in the gravitational field at $R$. But light propagates in the radial direction in the static field according to

$$d\tau^2 = 0 = g_{00}(r)dt^2 - g_{11}(r)dr^2 . \tag{4.56}$$

Therefore,

$$\Delta t = t_\infty - t_R = \int_R^\infty \left(\frac{g_{11}(r)}{g_{00}(r)}\right)^{1/2} dr \tag{4.57}$$

is the constant transit time for each wave crest to travel from $R$ to $\infty$. Hence the coordinate interval $dx^0$ at $R$ governed by (4.55) is also the coordinate interval between wave crests arriving at $\infty$ from $R$. Therefore, the stationary observer with his apparatus at the location denoted by $\infty$ measures the proper time interval between crests arriving from the star,

$$d\tau_{a'} = \sqrt{g_{00}(\infty)}\, dx^0 \tag{4.58}$$

[say by a process analogous to the inverse of (4.53)].

Denote a spacetime event associated with the distant observer by $x'^\mu$. Similarly to (4.55), we have

$$d\tau_a = \sqrt{g_{00}(\infty)}\, dx'^0 , \tag{4.59}$$

as the equation that governs the coordinate time $dx'^0$ for emission from the like atom at rest at the location of the observer, because (4.53) holds equally at both locations. So we can compare the *observed* frequency $w_o$ of the photon arriving from the surface $R$ of the star, as measured by the observer at $\infty$ with that of the photon *emitted* by the like atom located with the observer, namely, $w_e$. (Of course the atom at the location of the observer can be identified with that on the star by comparing the ratios of a few spectral lines.) Recall that we have proven above that both atoms will emit the same frequency photon as seen by a local observer near the atom. The proper time between crests (multiplied by $c$) gives the wavelength of the light or the reciprocal of the frequency. Therefore, using (4.58) and (4.59), respectively, the ratio of the frequency of light coming form the star and *observed* by the distant observer, to the frequency of light *emitted* by his local atom is

$$\frac{w_o}{w_e} = \frac{d\tau_a}{d\tau_{a'}} = \frac{dx'^0}{dx^0} = \left(\frac{g_{00}(R)}{g_{00}(\infty)}\right)^{1/2} . \tag{4.60}$$

In the last step, we eliminated $d\tau_a$ between (4.55) and (4.59). Thus, as observed at $\infty$, the photon arriving from the atom located on the surface of the star has a lower frequency

$$w_o = e^{\nu(R)} w_e = \sqrt{1 - 2M/R}\, w_e \tag{4.61}$$

than the photon from the identical atom located at $\infty$. The light is red-shifted. Conversely, if a photon fell from $\infty$ to the surface of the star and was observed there and compared with a photon from a like atom on the star surface, it would be found that the photon from the distant location had been blue-shifted.

We have introduced the Schwarzschild metric valid for a static spacetime outside a spherical star into the above equations. Otherwise we could keep the expressions in terms of the general metric functions. We can also write (4.61) in terms of energies. Thus, the energy observed at infinity is related to that of an event at $r$ by

$$E = E(r)\sqrt{g_{00}(r)} \qquad (4.62)$$

for any $r$.

The result above can be stated in terms of clocks: time flows more slowly at $R$ than at $\infty$. For an isolated static spherical star, the gravitational time dilation factor is $e^{-\nu(R)}$.

The conventionally defined gravitational redshift is the fractional change between observed and emitted wavelengths compared to emitted wavelength:

$$z \equiv \frac{\Delta\lambda}{\lambda_e} = \frac{\lambda_o}{\lambda_e} - 1 = \frac{\omega_e}{\omega_o} - 1\,, \qquad (4.63)$$

or, in the case of a Schwarzschild star,

$$z = \left(1 - \frac{2M}{R}\right)^{-1/2} - 1 \le \frac{M}{R}\,. \qquad (4.64)$$

We can calculate the limit to the redshift from the surface of a neutron star. From the Oppenheimer–Volkoff equations we have noted that $M/R < 1/2$. The equality yields the horizon. But relativistic stars actually observe a more stringent condition, $2M/R < 8/9$, for any star obeying the Oppenheimer–Volkoff equations, as will be shown in Section 4.16. This gives $z < 4/9$ as the maximum gravitational redshift of a stable star. In Fig. 4.7 we show the gravitational redshift of light emitted at the surface of neutron stars belonging to several model sequences.

The maximum gravitational redshift of light emitted from an atom on the surface of a stable star is enormously different from the redshift of light emitted from just outside the horizon of a black hole. For a star, the redshift is $z < 4/9$; for a black hole it tends to infinity as the atom approaches the Schwarzschild radius, as can be seen from (4.64).

We can apply (4.60) to the situation discussed in Section 3.1.1, that of a photon climbing up the constant weak gravitational field of the earth over a limited height $h$. The acceleration due to gravity is nearly constant over the limited height. The Newtonian potential is $V(r) = -M/r$ and its

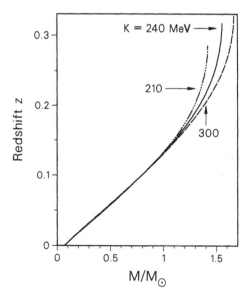

FIGURE 4.7. Surface gravitational redshift of several model neutron star sequences. The resistance of matter to compression is labeled by the value of $K$—the compression modulus. See Compact Stars [4] for more details.

relationship to $g_{00}$ was derived in Section 3.2.5:

$$\frac{\omega(R)}{\omega(R+h)} = \left(\frac{g_{00}(R)}{g_{00}(R+h)}\right)^{1/2} \approx 1 - V(R+h) + V(R)$$

$$\approx 1 - \frac{M}{R^2}h = 1 - gh. \tag{4.65}$$

We see that the frequency of the photon emitted from the atom on the surface of the earth $R$ compared at $R+h$ with a photon from a like atom at that location is

$$\omega(R) = (1 - gh)\,\omega(R+h), \tag{4.66}$$

in agreement with the earlier weak field result. Einstein discovered the approximate redshift formula (4.65) in 1911 before his discovery of General Relativity. He concluded that the local velocity of light is different according to the gravitational field it travels in. "In this theory the principle of constancy of light velocity does not apply in the same way as in ... the usual relativity theory." [74]

For Earth,

$$M_\oplus = 5.977 \times 10^{27}\,\text{g} = 0.4438\,\text{cm}, \tag{4.67}$$
$$R_\oplus = 6.371 \times 10^8\,\text{cm}, \tag{4.68}$$

so that

$$g_\oplus = \frac{M}{R^2} = 1.1 \times 10^{-18} \text{ cm}^{-1}. \qquad (4.69)$$

We see that the Newtonian approximation to the gravitational field is well justified on Earth.[12]

### 4.8.3    COMPARISON OF EMITTED AND RECEIVED LIGHT

The derivation above was carried out from the point of view of a hypothetical experiment: comparison of light received from a distant star with light from a laboratory source. Frequently, a simpler derivation of the redshift is given which simply compares (as the theorist can do) the frequencies of light emitted from an atom on the surface of the star and its frequency as received by a distant observer. The proper time corresponding to adjacent wave crests on the star was given by (4.55) which we now refer to with the subscript "e" for emitted,

$$d\tau_e = \sqrt{g_{00}(R)}dx^0. \qquad (4.70)$$

As shown above, the coordinate time taken for each wave crest to propagate from the emission location to the observer's location is a constant in a static field. So the observer measures the proper time

$$d\tau_o = \sqrt{g_{00}(\infty)}\,dx^0. \qquad (4.71)$$

The inverse is proportional to the frequency, so the ratio of the observed to emitted frequencies is

$$\frac{\omega_o}{\omega_e} = \left(\frac{g_{00}(R)}{g_{00}(\infty)}\right)^{1/2}, \qquad (4.72)$$

in agreement with the earlier result.

### 4.8.4    MEASUREMENTS OF $M/R$ FROM REDSHIFT

The potential usefulness of redshift measurements for compact stars is immediately evident from (4.61) in that it permits a determination of $M/R$. If either $M$ or $R$ is known from other measurements, say the mass from observations on a binary system, then the other quantity is known. So far, an independent measurement of, say $M$, has not been made for which a redshift measurement has also been made. Of course in comparing the two

---

[12]Note how helpful the appropriate choice of gravitational units was in the estimate. Mass and radius were both expressed in units of length so that acceleration was expressed in inverse length and when multiplied by height $h$ yielded a dimensionless number as needed in (4.66).

frequencies, one from the distant star, the other from an earth-bound atom, one cannot à priori be sure one is comparing like atoms. The identification would be more positive if several transition lines were observed and they all had the same ratio when compared to the atom on Earth. There are about 30 known X-ray pulsars for which tentative spectral identifications have been made.

## 4.9    White Dwarfs and Neutron Stars

"Chandrasekhar ... has shown that a star of a mass greater than a certain limit remains a perfect gas.... The star has to go on radiating and radiating, and contracting and contracting until, I suppose ... gravity becomes strong enough to hold in the radiation, and the star can at last find peace [as a black hole].

I felt driven to the conclusion that this was almost a *reductio ad absurdum*.... I think there should be a law of nature to prevent a star from behaving in this absurd way."
A. Eddington, 1935 [75]

"It is clear from this [above] statement that Eddington fully realized, already in 1935, that given the existence of an upper limit to the mass of degenerate configurations, one must contemplate the possibility of gravitational collapse leading to the formation of what we now call black holes. But he was unwilling to accept a conclusion that he so presciently drew."
S. Chandrasekhar, 1982 [76]

### 4.9.1    OVERVIEW

After sufficient time, a star, either a white dwarf or neutron star, will become cold on the nuclear scale. Ideally, it will be in the lowest energy state, and the neutrinos and photons produced by the reactions in achieving the lowest state, such as

$$
\begin{aligned}
e^- + p &\to n + \nu, \\
n &\to p + e^- + \bar{\nu}, \\
\bar{\nu} + \nu &\to 2\gamma,
\end{aligned}
\tag{4.73}
$$

will have escaped from the star. After the star has dropped in temperature below an MeV (= $1.1605 \times 10^{10}$ K), we may say that it is cold for the purpose of computing the energy and pressure, but the temperature is still very high compared to the neutrino masses (presumably very small or zero), so they have velocity sufficient to escape, if not light velocity. Typical white dwarf surface temperatures are a few times $10^4$ K with some as high as $10^6$

K. Interior temperatures are $10^6$ to $10^7$ K. Neutron stars are born very hot, of the order of $10^{10}$ K, but cool rapidly, reaching $10^8$ K in a month and $10^6$ K in less than a million years. White dwarfs have radii typically of a few thousand kilometers, compared to neutron stars which have radii of around 10 km. White dwarfs are about the smallest faintest stars that are visible in the optical spectrum. Neutron stars make their presence known in the radio spectrum as pulsars, highly magnetized rotating bodies.

Because of their low luminosity (generally several orders of magnitude less than the Sun's) known white dwarfs lie in the immediate neighborhood of the Sun at distances of 100 pc = 330 ly = $3.08 \times 10^{15}$ km. By comparison, the disk diameter of the Milky Way galaxy is $30,000$ pc = $9,900,000$ ly $\approx 10^{20}$ km.

White dwarf masses are sharply peaked around $0.6M_\odot$ though Sirius B has a fairly well determined mass $M = 1.053 \pm 0.028M_\odot$. The largest known mass is $1.52M_\odot$ (WD 1143+321), and the smallest is $0.33M_\odot$ (WD 0349+27). The latter are deduced from gravitational redshifts [77] and are less certain.

Many of the known pulsars are distributed around the Sun, close to the galactic plane. A few are known at distances of more than 15 kpc. It seems very likely that the proximity of known pulsars in the Galaxy is merely a problem of detecting their radio signals. In recent years pulsars have been discovered in globular clusters in which there seems to be an abundance; As of January 2006, 129 have been found in such clusters.

Few mass measurements have been made, and only four have high accuracy. The some 20 presently known masses of radio pulsars have their maximum likelihood values in the range 1.2 to 1.7 $M_\odot$, but the error bars would admit pulsars of masses 0.8 to 2.2 $M_\odot$ [56]. However, the binary system pulsars all have masses in the range $1.35 \pm .04M_\odot$; this narrow range as compared to the other measurements could be interpreted as a peculiarity in the formation of binary pulsars or because of the small error bars, a more accurate determination than the others.

White dwarfs are born under relatively quiescent circumstances, the remnants of a nonexplosive evolution of fairly modest stars of mass up to seven or so solar masses, which comprise most of the stars in the Milky Way galaxy. What remains of the core after the bulk of the star has expanded into a planetary nebula becomes a white dwarf. Dwarf progenitor stars, because of their low mass (unlike the progenitors of neutron stars), have not completed the chain of fusion reactions up to the most bound nucleus, $^{56}$Fe. They are therefore composed of matter that is not in its lowest energy state. They are too cold to ignite the nuclear reactions that would bring them there. However, even when cold and crystalline, white dwarfs may evolve on a very long timescale through zero temperature nuclear reactions in which lattice vibrations yield a small but finite probability of Coulomb barrier tunneling. Such a process known as a pycnonuclear reaction, produces some heavier elements [78, 79]. However, the composition is essentially determined at birth.

Because of the circumstances of their birth, white dwarfs are not members of a single sequence corresponding to a unique equation of state. Rather, according to its individual history, each white dwarf is an equilibrium stellar configuration corresponding to an equation of state describing the particular composition attained in the thermonuclear burning of its progenitor (mostly He, C, and O) together with an atmosphere that is the remnant of the vibrations of the envelope that became unstable and formed the planetary nebula.

On the other hand, neutron stars are born in enormous explosions, called supernovae, that occur in the last few moments in the evolution of stars more massive than about $8M_\odot$. In their cores, thermonuclear combustion has proceeded through to iron. Neutron stars are therefore composed of matter that has been fully processed to extract all available energy. At each density in their interiors, matter is in its ground state, consistent with charge neutrality. Therefore, neutron stars form a single sequence.

## 4.9.2    FERMI-GAS EQUATION OF STATE FOR NUCLEONS AND ELECTRONS

As an elementary exercise in the calculation of an equation of state appropriate for the description of compact stars, we consider an idealized composition of dense matter that will illustrate some of the relationships and contrasts between white dwarfs and neutron stars. We will also learn some of the principles involved in any calculation of the structure of compact stars. In the idealization, both types of stars are members of a single sequence belonging to a particularly simple equation of state. Oppenheimer and Volkoff adopted the equation of state of an ideal gas of neutrons for the first calculation of a neutron star model. However, pure neutron matter is unstable. Neutrons obey the Pauli exclusion principle so that some neutrons in dense matter would have more than enough energy to beta decay to proton, electron and neutrino. The neutrinos diffuse out of the star thus lowering the energy. On the other hand, the low-energy neutrons cannot decay because the proton and electron states are already occupied. We are interested in the equilibrium configuration of matter and of the star after all such processes have occurred and matter has reached its ground state at each relevant baryon density. Therefore neutrinos must be ignored in searching for the ground state of stellar matter.

Our first model for the equation of state of neutron stars and white dwarfs describes a gas of noninteracting neutrons, protons, and electrons in such proportions at each baryon number density that the gas has its lowest possible energy. Such a situation is referred to as beta equilibrium or simply equilibrium. The neutrons will not beta decay; nor will the inverse reaction—electron capture on a proton—occur. In addition, we must demand that the minimum energy be found subject to the constraint of electrical neutrality. Baryon density is usually employed as an independent variable in calculating an equation of state because baryon number is conserved. Baryon species may be transformed from one to another so as to minimize the energy, but the total baryon charge remains unchanged. The example we will consider is a prototype for any more sophisticated model of the equation of state of dense charge-neutral matter.

Because neutrons, protons, and electrons are all Fermions (particles of half odd-integer spin), they obey the Pauli exclusion principle—not more than one Fermion can occupy a given quantum state. In this section we deal with a *degenerate* ideal Fermi gas. Ideal in this context means that interactions are ignored. Degenerate means that all quantum states up to a given energy, called the Fermi energy, are occupied. Therefore, in summing over the occupied states (which in the absence of interactions are momentum eigenstates), whether over the particle number, over the energies, or whatever, we need to sum or integrate over momentum states. From quantum mechanics we recall normalization of momentum states in a box of dimension $L$ (c.f. Ref. [80]), so that

$$\frac{1}{L^3} \sum_{\boldsymbol{k}} \cdots \longrightarrow \int \frac{d^3k}{(2\pi)^3} \cdots = \frac{1}{2\pi^2} \int_0^{k_F} k^2 \, dk \cdots . \qquad (4.74)$$

For degenerate systems all energy states are filled in order up to some maximum called the Fermi energy, or in the case that momentum eigenstates are used, up to the Fermi momentum. The subscript "$F$" is used to denote the Fermi momentum or energy, where needed to avoid ambiguity. The last equality in (4.74) holds provided the integrand depends only on the magnitude of the wave number $k = |\boldsymbol{k}|$, as is usually the case.

Particle momentum $p$ and wave number $k$ are related in the usual way: $p = \hbar k$. But because we set $\hbar = 1 = c$, we refer to $k$ as momentum. Its unit is inverse length.[13] In numerical calculations it is also advantageous to convert mass to inverse length by dividing by $\hbar c = 197.33$ MeV-fm, or alternately, multiplying $k$ by $\hbar c$ and using MeV as the unit.

The assumption of degeneracy is valid for low temperatures—much less than the Fermi energy

$$T << E_F \equiv \sqrt{k_F^2 + m^2} \qquad \text{(degeneracy condition)} . \qquad (4.75)$$

Consider a white dwarf as an example. Obviously, a lower bound on the Fermi energy can be found by ignoring the momentum. For the electron Fermi energy we have $E_F > m_e = 0.511$ MeV $\sim 6 \times 10^9$ K. This is large

---

[13]One is accustomed to the combination $p^2c^2 + m^2c^4$ as particle energy squared. In natural units the expression will appear simply as $k^2 + m^2$. In computing a numerical value, $k$ and $m$ will have to be expressed in the same units, for example, as MeV or fm$^{-1}$.

compared to the internal temperature of most white dwarfs, $\sim 10^6$ to $10^7$ K. Thus, the Fermi energy of electrons (and therefore also nucleons) easily satisfies the degeneracy condition, and we will generally consider them to be cold stars in the above sense.

We wish to calculate the equation of state, which is the pressure $p$ and energy density $\epsilon$ as a function of number density $\rho$, or alternately, the value of the pressure at given energy density, $p(\epsilon)$. The number density is obviously obtained by summing over the occupied states in (4.74) and multiplying by the degeneracy of the momentum state. The energy is obtained by performing the same sum over $\sqrt{k^2 + m^2}$. Each Fermion type $(n, p, e)$ contributes to the energy density, pressure, and number density according to

$$\epsilon = \frac{\gamma}{2\pi^2} \int_0^k \sqrt{k^2 + m^2}\, k^2\, dk\,,$$

$$p = \frac{1}{3}\frac{\gamma}{2\pi^2} \int_0^k \frac{k^2}{\sqrt{k^2 + m^2}}\, k^2\, dk\,, \qquad (4.76)$$

$$\rho = \frac{\gamma}{2\pi^2} \int_0^k k^2 dk\,,$$

To verify that the pressure written above obeys the usual thermodynamic relationship,

$$p = -\frac{\partial E}{\partial V} = \rho^2 \frac{\partial}{\partial \rho}\frac{\epsilon}{\rho} = \rho\frac{\partial \epsilon}{\partial \rho} - \epsilon\,, \qquad (4.77)$$

perform the differentiation using the standard integrals below. Note that the last equality yields

$$\frac{\partial \epsilon}{\partial \rho} = \frac{\epsilon + p}{\rho}\,. \qquad (4.78)$$

In the above equations, $\gamma$ denotes the degeneracy of each momentum state. The degeneracy is 2 for each Fermion type corresponding to the two spin projections $\pm 1/2$. The upper limit on the integrals stands for the Fermi momentum, sometimes denoted by $k_F$. For the simplicity of the appearance of formulae to follow, we call it simply $k$, or $k_p$ when we have to distinguish proton Fermi momentum from others. Clearly, Choosing a value of $k$ is equivalent to choosing the Fermion density.

According to the units just described, $\epsilon$ and $p$ can be conveniently computed in units $1/\text{fm}^4$. They can be converted to $\text{MeV}/\text{fm}^3$ by multiplying by $\hbar c$ or still other units by using the conversion factors described in Section 4.3.

The integrals can be carried out in closed form using the results of Ref. [81], pp. 86 and 87:

$$\int_0^k \sqrt{k^2 + m^2}\, k^2\, dk =$$

$$\frac{1}{4}\left[k(k^2+m^2)^{3/2} - \frac{1}{2}m^2k\sqrt{k^2+m^2} - \frac{1}{2}m^4\ln\left(\frac{\sqrt{k^2+m^2}+k}{m}\right)\right],$$

$$\int_0^k \frac{k^4}{\sqrt{k^2+m^2}}dk =$$

$$\frac{1}{4}\left[k^3\sqrt{k^2+m^2} - \frac{3}{2}m^2k\sqrt{k^2+m^2} + \frac{3}{2}m^4\ln\left(\frac{\sqrt{k^2+m^2}+k}{m}\right)\right],$$

$$\int_0^k \frac{k^2}{\sqrt{k^2+m^2}}dk = \frac{1}{2}\left[k\sqrt{k^2+m^2} - m^2\ln\left(\frac{\sqrt{k^2+m^2}+k}{m}\right)\right].$$

As the contribution of each Fermion type (of degeneracy $\gamma = 2$), we find

$$\epsilon = \frac{1}{4\pi^2}\left[\mu k\left(\mu^2 - \frac{1}{2}m^2\right) - \frac{1}{2}m^4\ln\left(\frac{\mu+k}{m}\right)\right],$$

$$p = \frac{1}{12\pi^2}\left[\mu k\left(\mu^2 - \frac{5}{2}m^2\right) + \frac{3}{2}m^4\ln\left(\frac{\mu+k}{m}\right)\right], \qquad (4.79)$$

$$\rho = \frac{k^3}{3\pi^2},$$

where

$$\mu = (m^2 + k^2)^{1/2} \qquad (4.80)$$

is the Fermi energy (sometimes referred to as the chemical potential) and $k$ the Fermi momentum. We could eliminate $k$ in favor of $\rho$ in the above equations, yielding $\epsilon(\rho)$ and $p(\rho)$.

Continuing our calculation of the equation of state, we want to minimize the total energy density of neutrons, protons, and electrons $\epsilon(\rho_n, \rho_p, \rho_e) = \epsilon(\rho_n) + \epsilon(\rho_p) + \epsilon(\rho_e)$ at fixed baryon density $\rho = \rho_n + \rho_p$ and subject to the condition of charge neutrality $\rho_p = \rho_e$. This can be done by the method of Lagrange multipliers [82]. Construct a new function from the one we wish to extremize together with the two constraint equations:

$$F(\rho_n, \rho_p, \rho_n) \equiv \epsilon(\rho_n, \rho_p, \rho_n) + \alpha(\rho - \rho_n - \rho_p) + \beta(\rho_e - \rho_p).$$

For arbitrary variations of the three particle densities we require that

$$\frac{\partial F}{\partial \rho_n} = 0, \qquad \frac{\partial F}{\partial \rho_p} = 0, \qquad \frac{\partial F}{\partial \rho_e} = 0.$$

Using the integral expressions for $\epsilon$, we have

$$\frac{\partial \epsilon}{\partial \rho_n} = \frac{\partial \epsilon}{\partial k_n}\frac{\partial k_n}{\partial \rho_n} = \sqrt{k_n^2 + m_n^2}, \qquad \text{etc.} \qquad (4.81)$$

Therefore the above conditions on $F$ yield

$$
\begin{aligned}
\alpha &= \sqrt{k_n^2 + m_n^2} \equiv \mu_n, \\
\alpha + \beta &= \sqrt{k_p^2 + m_p^2} \equiv \mu_p, \\
-\beta &= \sqrt{k_e^2 + m_e^2} \equiv \mu_e.
\end{aligned}
\tag{4.82}
$$

Eliminating the Lagrange multipliers, we obtain

$$
\mu_e + \mu_p = \mu_n.
\tag{4.83}
$$

Because $\mu_e$ is the electron chemical potential, it corresponds to unit negative electric charge while $\mu_n$ corresponds to unit baryon charge without electric charge (i.e., neutron).

The expression of chemical equilibrium (4.83) will recur in more general terms many times. It ensures that the particle levels are filled to such a point that no energy can be extracted from the gas by a neutron undergoing a beta decay or a proton an inverse beta decay. Of course, all levels are Fermi-blocked below the Fermi energies.

It is particularly noteworthy that the chemical potentials of three types of particles can all be expressed in terms of the two (Lagrange multipliers) associated with the conservation laws, namely, baryon number and electric charge conservation. The system is said to have three components, of which only two are independent.

As an aside, we remark that the conclusion concerning the possibility of expressing the chemical potentials for $m$ substances in equilibrium in terms of the chemical potentials of the $n$ independent components (or as we shall often say, conserved charges) is not restricted to free particles as in the above example. In general the problem is to find the minimum of, say, the Gibbs free energy $\Phi$, subject to the subsidiary conditions that express the conservation of the independent components (charges). The problem is therefore of the same kind as solved above. The simple expressions for the chemical potentials (4.80) will not hold. But we will still have $\mu_i = (\partial \Phi / \partial N_i)_{P,T}$, and the derivation will go through as above.

There is an alternate way of deriving the relation of the chemical potentials for the $m$ substances in terms of those for the independent components. It consists of writing down all of the expressions for the possible chemical transformations in the form $\nu_i A_i = 0$, where the $\nu_i$ are positive or negative (and not necessarily integers, as in the case of quarks) and the $A_i$ are the chemical symbols. One then replaces the chemical symbols by the chemical potentials. If several chemical transformations are possible, there is an equation for each and the resulting relationships between the chemical potentials allows one to express the chemical potentials of all components in terms of the independent ones alone (see Chapter X of Ref. [83]).

Pursuing our search for the minimum of the total energy at fixed density subject to charge neutrality, we note that the Fermi momenta of nucleons and the baryon density $\rho$ are related by

$$\frac{1}{3\pi^2}(k_n^3 + k_p^3) = \rho.\tag{4.84}$$

The condition of charge neutrality can be written

$$k_p = k_e.\tag{4.85}$$

These two equations together with (4.83) can be used to determine the $k_n, k_p$, and $k_e$ such that the energy is minimized subject to the stated conditions. We now analyze how they may be solved.

Suppose $k_p = 0$. Then we obtain

$$k_n^2 = (m_p + m_e)^2 - m_n^2 < 0,\tag{4.86}$$

where the values of neutron and proton masses were written earlier and $m_e = 0.511$ MeV. So there is no real solution, and we conclude that $k_p \geq 0$. Next try $k_n = 0$. We find

$$k_p^2 = \left(\frac{m_e^2 + m_n^2 - m_p^2}{2m_n}\right)^2 - m_e^2 = 1.40 \text{ MeV}^2$$
$$= 3.60 \times 10^{-5} \text{ fm}^{-2}.\tag{4.87}$$

This is the smallest value of $k_p$ for which $k_n$ can be finite. Below the density

$$\rho = \frac{k_p^3}{3\pi^2} = 7.29 \times 10^{-9} \text{ fm}^{-3}\tag{4.88}$$

or below the energy density[14]

$$\epsilon \sim \rho m_p = 3.47 \times 10^{-8} \text{ fm}^{-4} = 1.22 \times 10^7 \text{ g/cm}^3,\tag{4.89}$$

a charge-neutral Fermi gas in equilibrium is an equal mixture of protons and electrons with no neutrons. Below this neutron threshold, we have simply

$$k_p = (3\pi^2\rho)^{1/3}, \quad k_e = k_p, \quad k_n = 0,\tag{4.90}$$

and the energy density and pressure can be computed from (4.79) or the low–density approximations derived in the next section.

Above the neutron threshold, we can use the equation for fixed baryon density to write

$$k_n = (3\pi^2\rho - k_p^3)^{1/3}\tag{4.91}$$

---

[14]Note that we are making frequent use of the conversion factors derived earlier in the chapter.

together with the condition of charge neutrality to obtain an equation in $k_p$ alone:

$$(k_p^2 + m_e^2)^{1/2} + (k_p^2 + m_p^2)^{1/2} = \left[(3\pi^2 \rho - k_p^3)^{2/3} + m_n^2\right]^{1/2}. \qquad (4.92)$$

This can be solved numerically for $k_p$ for any $\rho$, and then the corresponding $k_n$ can be found. The total energy density and pressure can then be calculated.

It is of some interest to look at the above equation in the ultrarelativistic regime where we ignore all masses. Then one obtains

$$\rho_p \rightarrow \frac{1}{8}\rho. \qquad (4.93)$$

If the existence of all other baryon types and their mutual interactions are ignored, the asymptotic proton density is one-eighth of the baryon density. One can also easily show in the same limit that this fraction is approached from below. For this reason it is sometimes thought that neutron stars are composed mostly of neutrons. Such an assumption is frequently used in calculations of neutron star properties, for example, masses, transport coefficients, or cooling rates.

The Fermi gas model for an equation of state has introduced some of the important principles involved in dealing with dense matter—the equilibrium composition of matter, charge neutrality, and the role of the Pauli principle. The directions in which the model can be improved, include (1) the introduction of additional baryon species such as hyperons, (2) nuclear interactions, (3) an important property of nuclear matter, namely, the energetic preference for an equal number of neutrons and protons, modulo effects of the Coulomb force. This is often referred to as the preference for *isospin symmetry*, about which we elaborate in Compact Stars [4], and finally (4) phase transitions, such as quark deconfinement or kaon condensation. So, within the overriding constraint of charge neutrality, we anticipate that the composition of neutron stars will be very rich in baryon species [84, 85]. We discuss these important topics in [4].

While the above formulae for energy density and pressure (4.79) are exact expressions for the equation of state within the limitations of the Fermi gas model, they are a poor basis for numerical computation in the case that a particle momentum is very small compared to its mass. In this case we must derive the nonrelativistic limits of the exact expressions and use them to compute the contributions to the energy density and pressure.

## 4.9.3  HIGH AND LOW–DENSITY LIMITS

For both the numerical reason mentioned above and for the sake of an interesting partially analytic discussion of white dwarfs as polytropes (which we shortly define), we derive the high- and low–density limits of the exact

Fermi gas expressions (4.79) and (4.76), respectively. The high–density limit $k >> m$, which may also be called the relativistic limit if $m$ is not ignored or the ultrarelativistic limit if it can be ignored. The results are

$$\epsilon \approx \frac{1}{4\pi^2}\left[k^4 - \tfrac{1}{2}m^4 \ln \frac{2k}{m}\right],$$

$$p \approx \frac{1}{12\pi^2}\left[k^4 + \tfrac{3}{2}m^4 \ln \frac{2k}{m}\right].$$

(4.94)

The logarithmic terms are small compared to $k^4$ in the ultrarelativistic limit, so we have

$$\epsilon \to 3p \approx \frac{1}{4\pi^2}\,(3\pi^2\rho)^{4/3} \quad \text{(high density, } k >> m),$$

(4.95)

where $\rho$ here is the density of the Fermion type considered.

The low–density expansion can easily be found directly from the integral expressions by expanding in $k/m$:

$$\epsilon \approx \frac{m^4}{\pi^2}\left[\frac{1}{3}\left(\frac{k}{m}\right)^3 + \frac{1}{10}\left(\frac{k}{m}\right)^5 - \frac{1}{56}\left(\frac{k}{m}\right)^7 + \frac{1}{144}\left(\frac{k}{m}\right)^9\right],$$

$$p \approx \frac{m^4}{3\pi^2}\left[\frac{1}{5}\left(\frac{k}{m}\right)^5 - \frac{1}{14}\left(\frac{k}{m}\right)^7 + \frac{1}{24}\left(\frac{k}{m}\right)^9\right].$$

For very low density, the nonrelativistic approximation involves the terms to $k^5$ only. When written in terms of $\rho$, they appear as

$$\epsilon \approx \rho m + \frac{(3\pi^2\rho)^{5/3}}{10\pi^2 m},$$

$$p \approx \frac{(3\pi^2\rho)^{5/3}}{15\pi^2 m} \quad \text{(low density, } k << m).$$

(4.96)

Notice that the term in $\epsilon$ proportional to $\rho$ does not contribute to the pressure. (See (4.77).) These are the contributions of each Fermion type to the equation of state.

Of course "low" and "high" density have different meanings for electrons and nucleons. For example, in the domain $m_e < k_e = k_p < m_p$, the electrons are relativistic ($k_e > m_e$), but the protons are nonrelativistic ($k_p < m_p$). This has to be kept in mind in computing an equation of state over a wide range of densities for $n, p$, and $e$ in equilibrium.

The equation of state of this ideal Fermi gas model is shown in Fig. 4.8. The kink at $\epsilon \approx 10^7$ gm/cm$^3$ marks the neutron threshold. Below this, the gas is an equal mixture of protons and electrons. Above, it rapidly becomes dominated by neutrons. The flat region in pressure corresponds simply to the fact that the increase in density in this region is accounted for by low-momentum (threshold) neutrons.

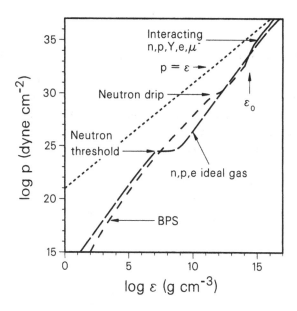

FIGURE 4.8. Various equations of state are compared with that of a charge-neutral Fermi gas equation of state of $n, p$, and $e$ in beta equilibrium, labeled by "ideal gas". Flat pressure region corresponds to the neutron threshold. For comparison, the so-called BPS equation of state (to which we will often refer) is also shown [86, 87, 88], which is continued at nuclear density by one that includes nucleons and hyperons [4]. (see also the Section on white dwarfs 4.10.) (Y denotes the hyperons.)

Another equation of state of greater relevance to neutron stars is also shown in Fig. 4.8. The BPS equation of state is a low-density equation of state that corresponds to the lowest energy state of nuclei embedded in a lattice with electrons in the interstices. The kink in this equation of state at $\epsilon \approx 4.3 \times 10^{11}$ gm/cm$^3$ is the density at which free neutrons drip from the nuclei. The scale up to about $\epsilon_0$ corresponds to low densities. Also shown is a high–density equation of state of hyperonized matter [4], namely the solid line above $\epsilon_0$. By "hyperonized" matter we mean that in addition to neutrons, protons, electrons and muons in a charge neutral equilibrium admixture, excited nucleon states and strange baryons are included. That some nucleons convert to hyperons in a neutron star is not forbidden by strangeness conservation, for that quantum number is conserved in strong interactions, but not on the weak interaction time scale of $\tau \sim 10^{-10}$ seconds.

The causal limit, $p = \epsilon$ is shown in Fig. 4.8 by the straight line. Any equation of state must lie below this limit, and can, at most, approach it asymptotically. The causal limit equation of state provides an extreme upper bound for the mass of neutron stars, $M \approx 3.2 M_\odot$.

### 4.9.4 POLYTROPES AND NEWTONIAN WHITE DWARFS

The results above for an ideal gas equation of state could serve to provide a crude model of white dwarf structure. However, what we have derived can be used for an improved description that recognizes the fact that white dwarfs do not consist of a gas of nucleons and electrons, but rather that the nucleons are clumped in nuclei, such as $^4$He, $^{12}$C, and $^{16}$O. (Recall page 73)

For charge neutrality, we require $\rho_p = \rho_e$ or, equivalently, $k_p = k_e$. Let us recall that charge-neutral matter consists only of protons and electrons below a certain nucleon density. The momentum corresponding to the neutron threshold is (4.87)

$$k_e = k_p = 0.00600 \text{ fm}^{-1} \quad \text{(neutron threshold)}. \tag{4.97}$$

This compares with the electron and proton masses of

$$
\begin{aligned}
m_e &= 0.511 \text{ MeV} = 0.511/197 \text{ fm}^{-1}, \\
&= 0.00259 \text{ fm}^{-1} \\
m_p &= 4.76 \text{ fm}^{-1},.
\end{aligned}
\tag{4.98}
$$
$$\tag{4.99}$$

Therefore, far below the neutron threshold, electrons are nonrelativistic. From (4.97) we find that the electron pressure is much greater than that of the protons,

$$\frac{p_e}{p_p} = \frac{m_p}{m_e} \gg 1 \quad (k_e \ll m_e). \tag{4.100}$$

Electrons are relativistic at and above the neutron threshold as seen above: $k_e > m_e$. Above the threshold up to the interesting condition that neutron and proton densities are equal (approximately so in nuclei), electrons are relativistic, and nucleons are nonrelativistic, $k_n, k_p \ll m_p$. For relativistic electrons refer to (4.97) and (4.95) to find

$$\frac{p_e}{p_p} = \frac{5}{4} \frac{m_p}{k_e} \gg 1 \quad (m_e \ll k_e \ll m_p). \tag{4.101}$$

The two equations above show that in the white dwarf domain defined by

$$k_p \text{ and } k_n \ll m_p \quad \text{(white dwarf domain)}, \tag{4.102}$$

electrons, whether relativistic or nonrelativistic, contribute virtually all the pressure, while the nucleons contribute virtually all the energy density. This is a fundamental fact of white dwarf physics. R. H. Fowler was the first to recognize that white dwarfs are supported against gravitational collapse by the pressure of degenerate electrons [89].

We discussed above the white dwarf domain up to the point where neutron and proton densities are about equal. However, because of the short-range attraction between nucleons, a dilute gas of nuclei is a lower energy state than a dilute gas of neutrons, protons, and electrons. Said another way, nucleons are bound in nuclei by about 8 MeV per nucleon so that the average mass of nucleons is less when in nuclei than free. We will return to this subject in a more exacting way.

In the meanwhile it is worth pursuing the present model by modifying the idealized results in a physically motivated ad hoc manner, namely, as a dilute gas of nuclei and electrons. This is straightforward because we have just seen that the nucleons do not contribute appreciably to the pressure. It is only their rest energy that is important. We can therefore consider that the nucleons are bound in nuclei. We express the energy density as a multiple of the nominal mass, $m_N$, of a nucleon bound in a nucleus times their number. We express the number of nucleons in the nucleus as a multiple

$$\nu = (\rho_p + \rho_n)/\rho_e \qquad (4.103)$$

times the electron density $\rho_e$. Thus, for the energy density we have

$$\epsilon = \rho_e m_N \nu. \qquad (4.104)$$

In the discussion of the preceding section, $\nu$ would have been determined by the equilibrium condition for a neutral Fermi gas of nucleons and electrons. In this section it is a parameter used to represent the nucleon–electron ratio of the dominant nuclear species, in particular, white dwarfs such as He, C, or O. For each of these, $\nu = 2$. We note that, from its definition, we can write

$$\rho_n = \rho_p(\nu - 1). \qquad (4.105)$$

Because the neutron threshold occurs at nonzero proton density, it must be so that $\nu \geq 1$.

In the white dwarf domain (4.102), it is easy to confirm that the electron pressure is always less than the electron energy density because of the results that can be deduced from the foregoing;

$$\frac{p_e}{\epsilon_e} \approx \frac{1}{5}\left(\frac{k_e}{m_e}\right)^2 \ll 1, \qquad k_e < m_e,$$

$$\qquad (4.106)$$

$$\frac{p_e}{\epsilon_e} \approx \frac{1}{3}, \qquad k_e > m_e.$$

The electron energy density itself is small compared to the rest mass density of the nucleons. Consequently, energy density dominates pressure, and we can write

$$p << \epsilon, \quad 4\pi r^3 p << M(r), \quad \frac{2M(r)}{r} << 1, \tag{4.107}$$

These inequalities assure that the Newtonian approximation to (4.31) is valid. The last of the inequalities follows from the fact that no region of a star can lie within its Schwarzschild radius [42]. It is also apparent that the pressure would not be monotonic unless the last inequality holds. Considering that the pressure at $r$ supports all material above $r$, pressure should be increasing monotonically from the edge to the center of the star. The middle inequality follows from the first: rearrange $M(r) \sim (4\pi/3)r^3\bar{\epsilon}$ to read $4\pi r^3 p \sim 3(p/\bar{\epsilon})M(r) << M(r)$.

After the Newtonian approximation has been made, divide what remains of (4.31) by $\epsilon$, differentiate by $r$, and combine with (4.32) to obtain

$$\frac{d}{dr}\left(\frac{r^2}{\epsilon}\frac{dp}{dr}\right) = -4\pi r^2 \epsilon. \tag{4.108}$$

This is the equation that governs the structure of a Newtonian star. As with the Oppenheimer–Volkoff structure equations, the Newtonian equation defines a one-parameter family of stars (sometimes referred to as a sequence) for any specified equation of state. The central energy density $\epsilon_c \equiv \epsilon(0)$ is a convenient means of parameterizing the continuum of stars belonging to the sequence.

We now derive a white dwarf sequence. The equation of state can be approximated by the following form in the white dwarf domain (4.102),

$$p = K\epsilon^\gamma. \tag{4.109}$$

A star whose equation of state is approximated by such a simple form is known as a *polytrope*. Polytropes are interesting as approximate models of stars because more than the usual amount of work can be done analytically. The power gamma in the polytrope is a constant which has two limiting values depending on whether the electrons are relativistic or nonrelativistic, as we will see in the following two sections.

Boundary conditions for Newtonian stars (4.108) can be found easily. Taking $r \to 0$ shows that, for finite $\epsilon(0)$, we must have $p'(0) = 0$. For a polytrope, $p' = \gamma(\epsilon'/\epsilon)p$ which shows that, for $\epsilon(0) \neq 0$, we must have $\epsilon'(0) = 0$. Summarizing, the boundary conditions on (4.108) are

$$\epsilon(0) = \epsilon_c, \quad \epsilon'(0) = 0. \tag{4.110}$$

By transforming variables from $r, \epsilon$ to $\xi, \theta$, we get a universal equation for polytropes parameterized by the polytropic index $\gamma$. The transformation is

$$r = \left(\frac{K\gamma}{4\pi(\gamma-1)}\right)^{1/2} \epsilon_c^{\frac{\gamma-2}{2}} \xi, \quad \epsilon = \epsilon_c \theta^{\frac{1}{\gamma-1}},$$

$$p = K\epsilon_c^{\gamma} \theta^{\frac{\gamma}{\gamma-1}}, \tag{4.111}$$

and the result is the *Lane–Emden* equation of index $(\gamma-1)^{-1} = n$,

$$\frac{1}{\xi^2}\frac{d}{d\xi}\xi^2\frac{d\theta}{d\xi} + \theta^{\frac{1}{\gamma-1}} = 0. \tag{4.112}$$

The boundary conditions corresponding to (4.110) are

$$\theta(0) = 1, \quad \theta'(0) = 0. \tag{4.113}$$

The Lane–Emden equation can be integrated numerically for various $\gamma$. It is easily verified that, for $\gamma > 6/5$, the Lane–Emden function $\theta(\xi)$ is monotonic decreasing with increasing argument (radius), as would be required of the pressure and energy density in a static star. The first zero occurs at a finite argument; call it $\xi_1$:

$$\theta(\xi_1) = 0. \tag{4.114}$$

Because this is the point where the pressure (4.111) has fallen monotonically to zero, $\xi_1$ marks the edge of the star. From the equations for the transformation of variables (4.111), the stellar radius is

$$R = \left(\frac{K\gamma}{4\pi(\gamma-1)}\right)^{1/2} \epsilon_c^{\frac{\gamma-2}{2}} \xi_1. \tag{4.115}$$

The mass is given by (4.32) as

$$\begin{aligned}
M &= \int_0^R 4\pi^2 \epsilon(r)\, dr \\
&= 4\pi\epsilon_c^{(3\gamma-4)/2}\left(\frac{K\gamma}{4\pi(\gamma-1)}\right)^{3/2}\int_0^{\xi_1}\xi^2\theta^{\frac{1}{\gamma-1}}\,d\xi. 
\end{aligned} \tag{4.116}$$

Using the Lane–Emden equation, the integral is easily evaluated:

$$\int_0^{\xi_1}\xi^2\theta^{\frac{1}{\gamma-1}}\,d\xi = -\int_0^{\xi_1}\frac{d}{d\xi}\xi^2\frac{d\theta}{d\xi}\,d\xi = -\xi_1^2\theta'(\xi_1). \tag{4.117}$$

Hence, the mass is found as

$$M = 4\pi\epsilon_c^{(3\gamma-4)/2}\left(\frac{K\gamma}{4\pi(\gamma-1)}\right)^{3/2}\xi_1^2|\theta'(\xi_1)|. \tag{4.118}$$

Thus, for a specific polytropic sequence, the mass and radius have the relationship

$$R \sim M^{(\gamma-2)/(3\gamma-4)}. \tag{4.119}$$

We return to these results after discussing the stability of stars in Section 4.19.

### 4.9.5  Nonrelativistic Electron Region

As noted above, the white dwarf region (4.102) itself can be discussed in terms of two regions according to whether $k_e$ is greater or less than the electron mass. In both cases the energy density is dominated by the rest-mass density of nucleons, whether they be free or bound in nuclei.

The electron pressure dominates over the proton pressure even in the region $k_p = k_e \ll m_e \ll m_p$. Neutrons are altogether absent in this region (in the Fermi gas model of the equation of state) because, as shown above, the value of $k_e$ at the neutron threshold is larger than $m_e$. So it is required that $\nu = 1$. From (4.104), we may write

$$k_e = \left(\frac{3\pi^2 \epsilon}{m_N \nu}\right)^{1/3}, \qquad (4.120)$$

from which we obtain from (4.97) an expression for the pressure as

$$p = K\epsilon^\gamma, \quad K = \frac{1}{15\pi^2 m_e}\left(\frac{3\pi^2}{m_N \nu}\right)^\gamma, \quad \gamma = 5/3. \qquad (4.121)$$

From the solution of (4.112), $\xi_1 = 3.65$ and $-\xi_1^2 \theta'(\xi_1) = 2.71$ for $\gamma = 5/3$, we find

$$M = 2.79\, \nu^{-2}\left(\frac{\epsilon_c}{\epsilon_k}\right)^{1/2} M_\odot \quad \text{(for } \gamma = 5/3), \qquad (4.122)$$

where $\epsilon_k = \rho_k m_N \nu$ and $\rho_k = m_e^3/3\pi^2$ are the densities corresponding to $k_e = m_e$. We cannot say precisely what the inequality $k_e \ll m_e$ implies about $\epsilon_c/\epsilon_k$, save that it is less than unity. However, we do understand that we are dealing with the lighter white dwarfs. A limit on the density of these stars can be estimated for the regime $k_e < m_e$. It is

$$\epsilon \approx m_p \frac{m_e^3}{3\pi^2} = 9.8 \times 10^5 \text{ gm/cm}^3. \qquad (4.123)$$

In the $\gamma = 5/3$ range of white dwarfs, the mass is an increasing function of central density, and the radius (4.115) is a decreasing function (for any $\gamma < 2$). A decrease of radius with an increase of mass is quite a general property of degenerate stars, and the physical reason is simple. The gravitational attraction grows as the mass increases, causing a greater compaction of the star.

### 4.9.6  Ultrarelativistic Electron Region: Asymptotic White Dwarf Mass

Now we study the ultrarelativistic region and the famous Chandrasekhar mass-limit for white dwarfs. The electron pressure dominates nucleon or nuclear pressures as noted above. We may say that the ponderous nuclei or nucleons provide mass but little pressure as compared to electrons which are light and fast. From (4.104) we find the electron density in terms of the total energy density, and substituting into (4.95) we find that the pressure is given by the polytropic form

$$p = K\epsilon^{\gamma}, \quad K = \frac{1}{12\pi^2}\left(\frac{3\pi^2}{m_N\nu}\right)^{\gamma}, \quad \gamma = 4/3. \tag{4.124}$$

This is the equation of state in the limit of ultrarelativistic electrons throughout the star. In this limit we see that the mass (4.118) is independent of central density and the radius (4.119) tends to zero as $\gamma \to 4/3$.

The value of the unique mass, given in terms of the solution of the Lane–Emden equation $\xi_1 = 6.90, \; -\xi_1^2\theta'(\xi_1) = 2.02$, for $\gamma = 4/3$ is composition-dependent and given by

$$M = 5.87\,\nu^{-2}M_{\odot}, \quad \text{for } \gamma = 4/3. \tag{4.125}$$

This is the famous Chandrasekhar mass limit for white dwarfs. To estimate its numerical value, we note that the region

$$m_e \ll k_e = k_p \ll m_p \tag{4.126}$$

is well above the neutron threshold. Therefore, we can contemplate approximately equal numbers of neutrons and protons (as in $^4$He, $^{12}$C, and $^{16}$O of which real white dwarfs are made). Then from (4.125) we find $M \approx 1.5M_{\odot}$. This is a rough estimate of the maximum white dwarf mass [17, 18].

We have thus learned that the sequence of white dwarfs beginning at low through to high density, or correspondingly from $k \ll m_e$ to $k \gg m_e$, has an equation of state in the two limits that are polytropes, and that the polytropic index falls from $\gamma = 5/3$ to $4/3$ along the sequence of increasing central density. This corresponds to a softening of the equation of state. Between the limits the equation of state can still be thought of as a polytrope but with $\gamma$ a function of density ranging between the limits.

It is important to notice that the polytropic index reaches the value $\gamma = 4/3$ only as all electrons in the star become ultrarelativistic. In this limit the radius is zero and the density is infinite. Along the sequence of dwarfs of increasing density, the radius decreases as the mass increases toward the limiting value (4.125). The limit obtained by Chandrasekhar is an asymptotic limit. However, the limit of real white dwarfs is of a different nature, as discussed in the next section. The limiting mass of real white dwarfs is reached at finite density.

Chandrasekhar, while still a student, discovered that completely degenerate configurations have a mass limit when one takes account of special relativity in computing the equation of state. Limits to stellar masses had never been known before. It is not clear that Chandrasekhar fully understood the implication of his limit by enquiring of the ultimate fate of a star with mass above the limit. Eddington did grasp the implication—the collapse to what we now call a black hole—but he dismissed it as absurd [75]. (See the quotations at the beginning of this section.)

Eddington was not alone in rejecting the notion of a black hole. Nor was he the first to conceive the notion. Some 150 years before Eddington's remark, the Rev. John Michell had conceived the possible existence of black holes—stars so dense that their light could not escape from them. However, the notion was forgotten and never stimulated further interest except possibly in Laplace (1795).

> "...supposing light to be attracted by the same force in proportion to its inertia, with other bodies, all light emitted from such a body would be made to return to it, by its own proper gravity. ...If there should really exist in nature any bodies, whose ...light could not arrive at us ...we could have no information from sight; yet, if any other luminous bodies should happen to revolve about them we might ...infer the existence of the central ones with some degree of probability .... "
>
> *John Michell. B.D, F.R.S.*, 1783 [from a letter dated November 27, 1783 to Henry Cavendish, and printed in the Philosophical Transactions of the Royal Society (London) 1793]

It was not until the work of Oppenheimer and Snyder (1939) on continued gravitational collapse that the notion of black holes reappeared, and, even at that, it was not until the mid-1960s that intense interest in such objects emerged. Indeed, even John Wheeler, until the mid 1950s, believed that other processes would intervene to prevent collapse of a star to a black hole. He was partially correct. Not all gravitational collapses result in the formation of black holes. Now we know that the fate of stars more massive than the Chandrasekhar limit is more varied and complicated than envisioned by Eddington. Much mass may be expelled in a planetary nebula or supernova. Even if the collapsing core is more massive than the limit for a white dwarf, the star may still be arrested in its collapse at the stage of a neutron star, and only if the collapsing object exceeds the limiting mass for neutron stars will it collapse forever *within* the Schwarzschild horizon. Shortly after Chandrasekhar's pioneering work, Landau advanced a general argument for a maximum mass that the Fermi pressure of a degenerate gas of relativistic particles can sustain against gravity [90]. Landau's argument is based on fundamental constants. The maximum possible mass is of the order of a solar mass for either degenerate electrons or neutrons (see Ref. [43] for details). Landau's limit, based on the Fermi pressure of relativistic particles of unnamed type, was therefore a prediction of the limiting mass of white dwarfs and neutron stars.

Landau employed the Newtonian gravitational potential, and for Newto-

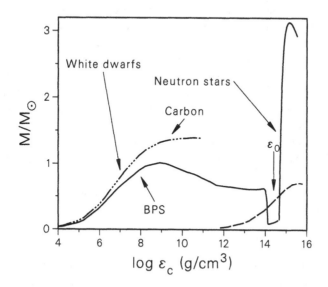

FIGURE 4.9. White dwarf sequences, one corresponding to the equation of state of BPS, the other to a pure carbon white dwarf (see Section 4.10) Two neutron star sequences are also shown: the solid line corresponds to the causal limit equation of state (see Section 4.18), and the dashed to the charge-neutral ideal $n, p, e$ gas in beta equilibrium. Stable stars have positive slope. Nuclear density is denoted by $\epsilon_0$.

nian gravity there is no limiting mass in the sense of General Relativity. The limit obtained by Chandrasekhar is also Newtonian and, as we have seen, is an asymptotic mass. Nevertheless, real white dwarfs do attain a mass limit along the sequence of equilibrium configurations, but at finite density, as depicted in Fig. 4.9, and it is customarily called the Chandrasekhar limit. We will discuss the nature of the limit in the next section.

Conditions for the validity of the Chandrasekhar limit—that all electrons are ultrarelativistic—can hardly be attained in real stars. After all, the pressure goes to zero at the stellar surface. However, it remains relevant to the iron core mass in stars near the termination of their evolution. In such stars, the weight of the overlaying matter can raise the Fermi energy of all electrons in the core. However, as we discuss next, capture of energetic electrons onto protons and nuclei still limits the Fermi energy in the core.

### 4.9.7 NATURE OF LIMITING MASS OF DWARFS AND NEUTRON STARS

We have seen that electrons become relativistic at a much lower density than nucleons and that the pressure of degenerate electrons—both non-relativistic and relativistic—establishes stable configurations called white dwarfs . These are Newtonian stars; general relativistic corrections are very small [91]. We also learned that the limiting mass of white dwarfs in the models that we have considered till now is attained only as an asymptotic limit at which the electrons are ultrarelativistic and the density is infinite.[15] This leaves little room for neutron stars in the sequence of degenerate stars.

The dilemma is resolved by noting that, even in the more realistic model considered above in which nucleons reside in nuclei (page 113), the neutron–proton ratio was held fixed. However, an important process is neglected in this approximation. As the electron Fermi energy increases, high-energy electrons are absorbed by protons—either free or in nuclei—in the inverse beta decay process,

$$^{A}Z + e^{-} \longrightarrow {}^{A}(Z-1) + \nu. \tag{4.127}$$

The neutrinos diffuse to the edge of the star, where, having the escape velocity, they depart. Therefore, the inverse reaction becomes impossible, and the star has reached its ground state (aside from thermal energy of its nuclear constituents). This process is also referred to as *neutronization*.

Neutronization depletes the supporting *electron pressure* along the sequence of stars parameterized by increasing mass. This brings the white dwarfs sequence to an end[16] at central densities of the order of $10^{9}$ g/cm$^3$. Therefore the mass limit for white dwarfs has to do with their changing internal composition [92, 93]. This limit is unrelated to the limit for relativistic stars, discussed below. What is at issue for white dwarfs is an instability occasioned by electron capture of energetic electrons. Consequently, the white dwarf mass limit is not quite the one envisioned by Chandrasekhar which, as previously noted, is an asymptotic limit attained only at infinite density.

To complete our discussion of the termination of the white dwarf sequence, we add that, if neutronization did not destabilize white dwarfs, the general relativistic instability discussed next would do so at a somewhat higher density [94].

White dwarf central densities range over about two orders of magnitude above and below $10^{7}$ gm/cm$^3$. At a much higher density, the Fermi pressure of nucleons (assisted by the short-range repulsion between them) establishes a second sequence of stable configurations, the neutron stars. Neutron star densities can be estimated by finding the neutron number

---

[15]In the limit $\gamma \longrightarrow 4/3$, we saw that the stellar radius goes to zero (4.119).

[16]See the section stability, 4.19 to understand that between white dwarfs and neutron stars there is a range of densities for there are no stable stars.

density corresponding to, say the point of transition from nonrelativistic to relativistic neutrons, $k_n = m_n$. In this case $\epsilon \sim \rho_n m_n \sim (1/(3\pi^2))m_n^4 = 6 \times 10^{15}$ gm/cm$^3$.

The neutronization instability of white dwarfs is in contrast to the gravitational collapse of a neutron star to a black hole or, indeed, the collapse of any relativistic star of sufficiently high mass concentration. There is no escape from collapse of any star having a sufficiently high concentration of mass near its center, even if made from hypothetical "incompressible" matter, as has been clearly argued by Harrison, Thorne, Wakano, and Wheeler [43]. They refer to the "no escape from collapse" in this connection.

Collapse of relativistic stars is intrinsic to the structure of Einstein's theory. This has to do with the fact that pressure, which in Newtonian stars resists collapse, in General Relativity makes collapse inevitable for sufficiently massive stars. This is because pressure appears together with energy density as a source of gravitational attraction in the Oppenheimer–Volkoff equation (3.180). (Recall Section 3.6.6.)

Indeed, in Section 4.16 we will see that even in the idealization that matter is incompressible, there is a limit to the mass and radius of a star: the limit depends simply on the assumed density of the incompressible medium! In general, compact relativistic stars have a mass limit that occurs at a finite central density. The particular value of the mass limit depends on the equation of state, but its existence does not.

## 4.9.8 DEGENERATE IDEAL GAS NEUTRON STAR

In our study of the Fermi gas equation of state for nucleons and electrons in beta equilibrium, we learned that at high baryon density the proton fraction approaches one-eighth from below. To first approximation, charge-neutral matter at densities above the white dwarf domain is pure neutron matter.

After the first few minutes of birth, a neutron star is very cold on the nuclear scale so that in the ideal gas model a neutron star is supported solely by the Fermi pressure of the degenerate neutrons, just as white dwarfs are supported by the pressure of degenerate electrons. Oppenheimer and Volkoff [3] were the first to investigate the structure of neutron stars, and they did so by assuming that the star was composed of noninteracting neutrons.

Oppenheimer and Volkoff found that their equations for the structure of relativistic stars yields a limiting mass of $\approx 0.72 M_\odot$ and a radius of $\approx 9.6$ km [3]. This is an interesting result for the following reason. We know that nuclear forces, especially the repulsive components, will become important at densities higher than the equilibrium density of nuclear matter. The nuclear force therefore provides additional resistance to gravitational collapse beyond that provided by the neutron Fermi pressure. So the Oppenheimer and Volkoff result establishes a value of the maximum mass of neutron stars that is close to but below the lower bound. The ideal gas of nucleons and electrons that we developed earlier in Section 4.9.2 is an even softer equation of state. It allows the pressure of neutrons at the top of their Fermi sea to be relieved by decay to protons and electrons (always in a charge-neutral equilibrium admixture). The limiting mass in this case is $0.7M_\odot$—even lower than the Oppenheimer and Volkoff lower bound.

For real neutron stars, the nuclear force is decisive in establishing both the limiting mass and how high it lies above the lower bound just discussed. In Section 4.18 we estimate the maximum amount by which the nuclear force can raise the mass limit above that established by the pressure of a degenerate ideal gas of nucleons and leptons.

The short-range repulsion of the nuclear force stiffens the equation of state compared to that of an ideal gas. Reference to stiff or soft with regard to an equation of state is a relative term. By "stiffer" it is meant that at given energy density the pressure is higher in comparison with a "softer" equation of state. The stiffest equation of state that is compatible with the notion of causality is one for which the speed of sound is the light speed.

However, limiting the sound speed to light speed does not uniquely define the equation of state because the sound speed is given by a derivative $\sqrt{dp/d\epsilon}$. This would allow any normalization of the equation of state. For the purpose of describing neutron stars, the equation of state should obviously match smoothly to a low–density equation of state that describes matter in a regime better known to us. This selects a normalization. We refer to an equation of state at the causal limit at high density and normalized in the above manner near nuclear saturation density as a *causal limit equation of state*. Such an equation of state provides the most resistance to collapse and therefore yields an upper bound on the limiting mass. Ruffini found the maximum possible mass of a nonrotating neutron star to be $\approx 3.2M_\odot$ (see Section 4.18). Figure 4.9 compares neutron star sequences for the two extremes of soft and stiff equations of state. The limiting mass of neutron star models based on realistic equations of state will lie between the two extremes.

Most nuclear models yield an the upper limit that lies considerably far from the two bounding models shown Figure 4.9. Nevertheless, the upper bound for the mass of neutron stars is very useful to astronomers for identifying celestial black holes. Invisible objects are sometimes detected by observation of the motion of a visible companion. Frequently the mass of each member of such a close binary pair can be inferred from a measurement of the orbital parameters. The invisible companion could be a neutron star that is either not emitting a radio beam, or the radio beam does not intersects our line of sight. In either case it is an invisible companion. Whether it is a neutron star or a black hole can be decided usually on the basis of a mass determination. If the mass is above the upper mass limit for neutron stars, then it is safe to infer that the unseen object is a black hole.

# 4.10   Improvements in White Dwarf Models

## 4.10.1   NATURE OF MATTER AT DWARF AND NEUTRON STAR DENSITIES

A number of the principles involved in the structure of compact stars have been encountered in the above simple model of a degenerate Fermi gas of free neutrons, protons, and electrons in equilibrium. The major objection that can be raised against it as far as white dwarfs are concerned, is that at densities below that of normal nuclei—and white dwarfs lie far below—it is energetically more favorable for nucleons to clump together into nuclei so as to exploit the short-range attractive nuclear force than to be uniformly dispersed at a lower density [43, 95, 86].

A white dwarf spans an enormous range of densities from its center to its surface. For example, Sirius B ($M = 1.053 M_\odot$, $R = 5,400$ km) has an average density of

$$
\begin{aligned}
\bar{\epsilon} &= \frac{M}{(4\pi/3)R^3} \\
&= \frac{1.053 \times 1.477 \text{ km}}{(4\pi/3)(5,400)^3 \text{ km}^3} \\
&= 2.35 \times 10^{-12} \text{ km}^{-2} = 3.17 \times 10^6 \text{ g/cm}^3 \\
&= 4.03 \times 10^5 \epsilon_{\text{Fe}} = 1.26 \times 10^{-8} \epsilon_0 ,
\end{aligned}
\tag{4.128}
$$

where nuclear saturation density is,

$$
\epsilon_0 = 2.51 \times 10^{14} .
\tag{4.129}
$$

Iron density is denoted by $\epsilon_{\text{Fe}}$. The central density is even higher than the average; the character of matter varies appreciably in this vast range of densities. For white dwarfs the constitution of matter is also very much influenced by the progenitor star and the extent of nuclear burning that preceded the formation of the white dwarf.

White dwarfs, unlike neutron stars, are not the remnants of progenitors in which nuclear fusion reactions have been driven to the iron end point under the relentless pressure of gravity (recall page 103). Nuclear combustion ceases in stars of a few solar masses at some intermediate element in the low-mass progenitors of dwarfs, so that dwarfs are composed mostly of $^4$He, $^{12}$C, or $^{16}$O in some mixture. While such a composition is not the lowest possible energy state, the lowest energy state is not always attainable in a finite time, even astronomical time. Therefore, the most favored state energetically is actually not achieved in white dwarfs. White dwarfs are much more individualistic (depending on the details of the evolution of the particular progenitor) than they would be if nuclear burning had proceeded through to iron, as in the cores of massive stars prior to their collapse to neutron stars or black holes [95, 96].

The pressure is zero at the edge of a star, and the energy density is low. Whatever elements are present at the surface, whether white dwarf or neutron star, are in their atomic state and are possibly ionized by the high temperature.

Neutron stars are formed from matter that has been fully processed by nuclear combustion, so that all available energy has been extracted at each density. Such hadronic matter in its ground state, is often referred to as "cold catalyzed matter" [43]. At zero pressure, such matter is composed of atoms of the most strongly bound nucleus, $^{56}$Fe arranged as in ordinary solid iron. At increasing density, the atoms become progressively more ionized, and the electrons fill the interstices. A Coulomb lattice arrangement of nuclei in the electron gas minimizes the energy. Matter at and near the surface of white dwarfs is similar except that the heaviest element is He, C, O, or Mg, depending on the degree of burning in the progenitor. As we mentioned already, white dwarfs, even of the same mass, are variable in composition, whereas neutron stars of the same mass should be identical.[17]

Whether neutron star or white dwarf, as the density increases from that at the stellar surface, electrons become increasingly relativistic. A lower energy state is achieved through the capture of energetic electrons by nuclei—inverse beta decay. Any neutrinos or photons produced diffuse out of the star, thus lowering its energy. With increasing density, nuclei become increasingly neutron-rich by this neutronization process. Neutronization sets in at a density for which the electron chemical potential, or Fermi energy $\mu_e \equiv \sqrt{m_e^2 + k_e^2}$, equals the threshold for the electron capture reaction (4.127).

To estimate the threshold for neutronization, let us take the proton-neutron mass difference $Q \sim -1$ MeV, as an approximate value. The threshold density for neutronization is therefore[18] Thus, from the above approximation, $k_e^2 = Q^2 - m_e^2$ we have

$$\rho = \frac{k_e^3}{3\pi^4} = \frac{(Q^2 - m_e^2)^{3/2}}{3\pi^2}, \qquad (4.130)$$

whence the electron or proton energy density $\epsilon = \rho \nu m_n$. Hence for the neutronization threshold we have,

$$\epsilon_{\text{neut}} \approx \frac{(Q^2 - m_e^2)^{3/2}}{3\pi^2} \nu m_N \sim 9 \times 10^6 \text{ gm/cm}^3. \qquad (4.131)$$

---

[17]However, for millisecond pulsars—those that rotate hundreds of times per second—the changing centrifugal effects as the star spins down significantly alter the distribution of density in the star with time. The star's internal constitution will reflect these changes, and the spin-down itself can be effected remarkably by the internal changes, as discussed in Ref. [4]

[18]For the estimate of the neutronization density we have taken the baryon electron ratio $\nu$ as 2 ,and $m_N$ as the typical energy of a nucleon in a nucleus, as 930 MeV.

Neutronization ultimately brings the white dwarf sequence to an end at a density of $\approx 10^9$ gm/cm$^3$ by robbing the star of its supporting electron pressure. For neutron stars, neutronization has only a minor effect, altering the preferred nuclear species in the crust from ordinary nuclei to neutron rich nuclei [86].

At higher density,

$$\epsilon_{\mathrm{drip}} \approx 4 \times 10^{11} \text{ gm/cm}^3 \,, \tag{4.132}$$

which is above the highest found in white dwarfs, but still low compared to the central density of neutron stars, another threshold is reached—neutron drip. The most weakly bound neutrons drip out of nuclei, and a gas of neutrons and electrons occupies the interstices in equilibrium with the nuclei [86].

At still much higher density, above the saturation density of nuclear matter (4.24)

$$\epsilon_0 = 2.51 \times 10^{14} \text{ gm/cm}^3 \,, \tag{4.133}$$

nuclei disassemble into a uniform charge-neutral mixture of baryons and leptons. This is the superdense regime, and with reference to neutron stars, is what we refer to as *neutron star matter*. We mean by this phrase that nuclear matter is in its lowest energy state at each density under the constraint of charge neutrality. Superdense matter contains as many baryon species at each density as is required for equilibrium so that no particle transformations by beta decay or otherwise will take place.

Hyperons—baryons with one or two strange quarks—form an important part of the population of superdense matter. In a later section and also in Compact Stars [4], we discuss these developments in our understanding of neutron star interiors, paying special attention to hyperon stars, and also, those whose core density lies above the critical density for deconfinement of quarks in individual baryons. Superdense matter, besides being rich in baryon species, may be composed of the baryon constituents—quarks—at high enough density.

The "low–density domain", below nuclear density, is not the subject of this book. But neutron stars span a density range from ordinary iron at the surface to superdense matter in the core. The crust region—the outer shell of a kilometer or less—is composed of ever more neutron-rich partially ionized to totally ionized atoms with greater depth in the crust. The ionized atoms are arranged on a Coulomb lattice in a gas of electrons. Above the drip density, neutrons also occupy the interstices between atoms as a diffuse gas. To compute stellar properties, even when dominated by the superdense regime, we must have an equation of state from the highest density down to the surface density. Therefore we make use of the extensive studies of the "low–density" regime discussed briefly below.

## 4.10.2  LOW–DENSITY EQUATION OF STATE

A number of authors have contributed to the calculation of the equation of state for cold catalyzed matter in various density domains below nuclear saturation: Feynman–Metropolis–Teller from 15 to $10^4$ grams/cm$^3$, Chandrasekhar upwards to $10^7$ grams/cm$^3$, Harrison–Wheeler–Wakano to $10^{12}$ grams/cm$^3$, Bonche–Negele–Vautherin to $10^{13}$ grams/cm$^3$. The compendium of results cited above are sometimes referred to as the HW equation of state [43].

Another compendium involving detailed calculation at subnuclear densities can be found in the work by Bethe–Baym–Pethick–Sutherland–Siemens [86, 87, 88], often referred to as the BPS equation of state. Both of these works, as remarked earlier, describe matter in its cold catalyzed state, that is, in the absolute ground state of charge–neutral hadronic matter. For both, the pressure becomes zero at the density of iron, the most bound state of low–density matter. The low–density equations of state are tabulated in ref. [4].

Both equations of state, the HW or BPS, are suitable for the description of the envelope of neutron stars, which are the product of progenitor stars sufficiently massive as to have sustained nuclear burning to the iron end point. Because zero pressure establishes the edge of a star, the surface of a neutron star is iron. A cold degenerate star has no atmosphere in the sense of the various gases that make up the earth's atmosphere.

The equations of state corresponding to cold catalyzed matter is not the best description of white dwarf material which is the product of stars whose burning stops at helium, carbon, or oxygen. As such, dwarfs do not belong to a single sequence of stars having the same equation of state. This is perhaps the inference to be drawn from Fig. 4.10 which shows radii and masses of observed white dwarfs [77] and shows a curve corresponding to the BPS equation of state representing matter in its ground state at each density. Also shown is a sequence of carbon white dwarfs which we discuss in the next section and which was qualitatively described above. The BPS equation of state yields a white dwarf mass limit of $\sim 1 M_\odot$ while a C white dwarf (closer to an accurate description of a real dwarf) has a limit $\sim 1.4 M_\odot$ (see Fig. 4.9).

Except in certain instances, the mass and radius "measurements" shown in Fig. 4.10 are highly model-dependent. Gravitational redshift measurements establish a relationship between mass and radius for a given star. The radius of a white dwarf is generally inferred from its luminosity and models of the stellar atmosphere, which presumably give a rough estimate but not likely an accurate one. Therefore both mass and radius depend on the atmospheric model.

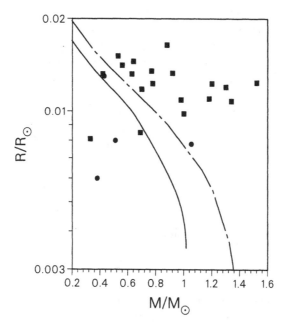

FIGURE 4.10. White dwarf radii and masses compared to the star sequence of the BPS equation of state (solid curve) representing equilibrium matter at each density. Also for carbon white dwarfs (dashed curve). However, see qualifications at the end of this section. (For comparison, recall that $R_\odot \approx 7 \times 10^5$ km.)

## 4.10.3   CARBON AND OXYGEN WHITE DWARFS

White dwarfs constitute the final form into which stars of a few solar masses evolve. Mass loss occurs during their evolution. Because of their low mass, thermonuclear reactions do not burn through to iron, but terminate at lighter nuclei, helium, carbon, oxygen, magnesium, or mixtures of these elements. An equation of state corresponding to fully catalyzed matter is therefore not quite an accurate description. Here we do a simple calculation of the equation of state of dwarfs that consist of a single nuclear species. More accurate calculations for such white dwarfs were done many years ago by Hamada and Salpeter [95].

Except at the very lowest densities of dwarfs, atoms are compressed to the point that nuclei are fully ionized. Therefore, to lowest approximation, we consider a fully ionized nuclear species immersed in a background of uniform density electrons. The positively charged nuclei will arrange themselves into a lattice with the nuclei at the lattice sites so as to minimize the Coulomb energy. We approximate this situation in the way suggested by Wigner and Seitz, that is, by considering a sphere surrounding each nucleus and touching the neighboring spheres. Each sphere containing a nucleus and its electrons will be neutral. To a good approximation there is no interaction between spheres. We consider one of them.

Most of the mass is contributed by the nuclei, whereas most of the pressure by the electrons (Section 4.9.4). The energy in each sphere consists of the nuclear mass, the electron mass and kinetic energy, the energy of Coulomb repulsion among the electrons, and the attractive Coulomb energy between the electron sea and the nuclei.

The Coulomb energy needed to assemble the electrons into the cell of radius $R$ is a familiar calculation. We need to integrate the interaction between the charge within the radius $r$ and the charge in a shell at $r$ of thickness $dr$,

$$E_{\text{self}} = e^2 \int_0^R \frac{1}{r}\left(\frac{4\pi r^3}{3}\rho_e\right)\cdot\left(4\pi r^2\, dr\, \rho_e\right) = \frac{3}{5}\frac{(Ze)^2}{R}, \tag{4.134}$$

where the electron number density is given by

$$Z = V\rho_e, \qquad V = \frac{4\pi}{3}R^3, \tag{4.135}$$

with $Z$ the atomic number of the nucleus and therefore the electron number per cell. The electron gas will have an attractive interaction with the nucleus at the center of the cell,

$$E_{\text{int}} = -\int \frac{1}{r}(Ze)(4\pi r^2\, dr\, e\rho_e) = -\frac{3}{2}\frac{(Ze)^2}{R}. \tag{4.136}$$

Adding these we get the lattice energy

$$E_{\text{lat}} = -\frac{9}{10}\frac{(Ze)^2}{R}. \tag{4.137}$$

Eliminating $R$ from the above expression in favor of $V$, we have the lattice contribution to the pressure,

$$p_{\text{lat}} = -\frac{\partial E_{\text{lat}}}{\partial V} = -\frac{3}{10}\left(\frac{4\pi}{3}\right)^{1/3} Z^{2/3}\, e^2 \rho_e^{4/3}. \tag{4.138}$$

To compute the equation of state in the present approximation for a pure nuclear species, we may choose a baryon density $\rho$ for which case the cell radius $R$ is given by the solution of

$$A = \frac{4}{3}\pi R^3 \rho, \tag{4.139}$$

where $A$ is the atomic number of the nuclear species (total nucleon number) and $Z$ its proton number. For such a chosen density $\rho$, the electron density is $\rho_e = (Z/A)\rho$. We treat the free electrons as a degenerate Fermi gas with maximum wave number

$$k_e = (3\pi^2 \rho_e)^{1/3} = \left(3\pi^2 \frac{Z}{A}\rho\right)^{1/3}, \tag{4.140}$$

and we denote the corresponding energy density and pressure by $\epsilon_e(k_e)$ and $p_e(k_e)$, respectively. We can compute these from (4.79, 4.95, 4.96) as the numerics dictate. The mass of the nuclei we take from experiment. These masses are measured typically as atomic masses, so we subtract the electron rest masses, but we ignore their atomic bindings. The energy density is $\epsilon = E_{\text{total}}/V$ with the volume $V = A/\rho$. So the energy density and pressure are

$$\epsilon(\rho) = \frac{\rho}{A}\left(M(A,Z) - Zm_e - \frac{9}{10}\frac{(Ze)^2}{R}\right) + \epsilon_e(k_e),$$

$$p(\rho) = p_e(k_e) - \frac{3}{10}\left(\frac{4\pi}{3}\right)^{1/3}Z^{2/3}e^2\rho_e^{4/3}. \qquad (4.141)$$

In the above, $M(A,Z)$ denotes the atomic mass. For example,

$$M(^4\text{He}) = 4.00260326U, \qquad M(^{12}\text{C}) = 12U,$$
$$M(^{16}\text{O}) = 15.99491502U, \qquad U = 931.504 \text{ MeV}. \qquad (4.142)$$

where $U$ is called the atomic mass unit, chosen (arbitrarily) so that $C^{12}$ is the standard. This is of no interest except when employing accurate mass tables.

The above equation of state cannot be valid above the density at which high-energy electrons are captured by protons in the nuclei. We have already discussed how the white dwarf sequences are terminated by a decrease of the supporting electron pressure because of inverse beta decay. To account for this process, one should evaluate the difference in energies (4.141) between a medium composed of any of the nuclear species (A,Z), (A−1,Z−1),..., (A,Z−1),... to find which is the lower and at what density.

To estimate the lower density range at which the approximation of fully ionized nuclei is valid, we may compare the Bohr radius of the inner orbit of the atomic species we have in mind with the cell radius which depends on the nucleon density, as in the relationship written above between $A$ and the volume of the cell. For simplicity, let $2Z \approx A$, which is exact in the case of He, C, and O stars. The two radii are

$$r_Z = \frac{r_{\text{Bohr}}}{Z} = \frac{\hbar^2}{m_e Z e^2}, \qquad R = \left(\frac{3Z}{2\pi\rho}\right)^{1/3}. \qquad (4.143)$$

The first is large compared to the second for baryon densities that satisfy

$$\rho > \frac{3Z^4}{2\pi}\frac{1}{r_{\text{Bohr}}^3} \qquad (4.144)$$

or energy densities

$$\epsilon > \rho U = 5.4Z^4 \text{ g/cm}^3. \qquad (4.145)$$

The computed range for which the fully ionized nuclei occupy Coulomb lattice sites holds for He, C, O, and Mg white dwarfs. Figure 4.10 shows the mass–radius relationship for carbon white dwarfs for an equation of state as computed in the manner described above. We note that the mass limits for the BPS and C equations of state are, respectively $\sim 1 M_\odot$ and $\sim 1.4 M_\odot$, which illustrates the dependence of gross properties of white dwarfs on their composition.

The data on mass and radius of observed white dwarfs lies mostly to the right of the region where the two different types of dwarf sequences lie. Two types of measurements are shown. The circles show measurements from binaries (giving $M$) combined with redshift measurements (giving $M/R$) and are rather accurate. The squares are probably much less certain. Again, they involve the redshift measurement, but the other independent determination, this time of radius, is obtained through the less certain chain involving color and flux, distances as measured by parallax, and model atmospheres [77, 97].

There appear to be problems of two kinds in the comparison with theory. To begin, we note that the two curves should rather accurately bracket the theoretical expectations for cold dwarfs. The BPS curve corresponds to an equation of state consisting of a Coulomb lattice of varying nuclear species, depending on density, embedded in an electron gas. The nuclear species and the arrangement in a lattice represents the lowest energy state of electrically neutral matter in the white dwarf domain. Such an equation of state should yield a sequence of dwarfs more compacted than can be realized in nature given the evolution of the progenitors that produce white dwarfs. The carbon equation of state represents a lattice of carbon nuclei with free electrons. It is appropriate to a white dwarf produced before the endpoint of nucleosynthesis is reached in the evolution of its progenitor.

As remarked earlier, essentially the same result for the equation of state is obtained for pure He and O stars as for C. The reason is simply that the result can be thought of as that of a polytrope with an electron to nucleon fraction of $1/2$, and with the polytropic index changing continuously from the nonrelativistic value $(5/3)$ to relativistic $(4/3)$ from low to high density. So the equations of state differ only in the small lattice energy, and pressure in going from He to O. The import of this note is that all cold white dwarfs should lie in the band between the two curves of Fig. 4.10. They clearly do not. As discussed, mass measurements for most white dwarfs are not directly made as for binaries, but depend strongly on model-dependent inferences about the radii. Because the equation of state is derived in a physically dependable regime, the inferred mass or radius determinations or both are most likely inaccurate.

## 4.11   Temperature and Neutron Star Surface

The temperature of neutron stars falls below 1 MeV in 20 seconds so that we may consider them as cold on the nuclear scale [51]. However, close to the surface where the density is low, say below the neutron drip density (4.132) and still closer to the surface than this, temperature becomes important, at least as a matter of principle. In practice, the mass, radius and other such bulk properties can be accurately computed as if the star is cold, and this is the assumption made in the computation of the high–density equation of state (justifiably) and also for low densities as in the work of Baym et al. and of Harrison and Wheeler [93, 86].

On the other hand, for such matters as cooling and phenomena associated with the magnetosphere, the zero temperature assumption for the surface layer becomes questionable. We have already noted that because a neutron star is made from the processed material of a massive star at the endpoint of its evolution, the most bound nucleus $Fe^{56}$ will be the dominant species at its surface. What is the nature of iron on and near the surface? We have noted that neutron stars are born with temperatures of about $10^{11}$ K and cool quickly to have surface temperatures of $10^{8}$ K in a month and $10^{6}$ K in less than a million years. The melting point of metallic iron at zero pressure is about 1,200 K and the boiling point is about 2,500 K. Certainly ordinary iron atoms cannot exist on the surface.

Let us define the surface region as that extending outward from the point where the density corresponds to the neutron drip density. At that density, the space available to each ion is far smaller than the size of atoms, as we compute shortly. Atoms are completely ionized. We will show that the lattice energy (4.137) is far larger than the thermal energy $kT$, so the lattice is secure against melting. However, close to the surface, where the density is lower, ionization by pressure becomes less important than thermal ionization. On the surface itself, simple estimates are impossible.

We make some crude estimates of conditions at several densities based on the preceding section by calculating the energy of a Coulomb lattice of iron *nuclei* (not atoms). First we consider matter at a depth within the star where the density is so high that atoms are fully ionized because there is insufficient room for atoms. Earlier we referred to the neutron drip density (page 125). It is a density occurring typically less than a kilometer from the surface, where some of the neutrons drip out of ionized atomic nuclei and join electrons as an interstitial gas. The energy density is given by (4.132). The corresponding baryon number density is

$$\rho_{\text{drip}} \approx 2.4 \times 10^{-4} \text{ fm}^{-3}, \tag{4.146}$$

obtained by dividing the energy density by the nucleon mass in grams. In the previous section we assumed that ionized atoms will form a Coulomb lattice. Here we want to compare the energy of the lattice with the thermal energy. From (4.143) we find the cell size corresponding to a Coulomb lattice of $_{26}Fe^{56}$ (actually approximated as an $A = 2Z$ nucleus for convenience),

$$R = \left( \frac{3 \times 26}{2\pi \times 2.4 \times 10^{-4}} \right)^{1/3} \text{ fm} = 37 \text{ fm}. \tag{4.147}$$

We now use this to calculate the lattice energy of fully ionized iron.

$$E_{\text{lat}} = \frac{9}{10} \frac{(26)^2 \times 1.44 \text{ MeV fm}}{37 \text{ fm}} = 24 \text{ MeV}. \tag{4.148}$$

Because $10^{10}$ K $\approx 1$ MeV, the temperature of the star expressed in MeV after a month is $10^{-2}$ MeV and in a million years $10^{-4}$ MeV. So the Coulomb lattice at the depth of the drip density is not disturbed by the temperature; neither is it disturbed at greater depth where a Coulomb lattice can exist to the density at which nuclei dissolve into their constituents.

Let us now enquire to what smaller density and lattice energy the temperature can also be ignored. To do this, let us find the density at which each nucleus is allotted a volume corresponding to the lowest Bohr orbit of iron. At that density the atoms must be fully ionized by virtue of the high *density*. Then compare the lattice energy with temperature. The Bohr radius for the lowest electron in iron is

$$r_Z = \frac{\hbar^2}{m_e Z e^2} = \frac{(\hbar c)^2}{m_e c^2 \cdot Z e^2} = 2 \times 10^3 \text{ fm}. \tag{4.149}$$

Find the number density at which the cell radius $R$ equals $r_Z$. It is

$$\rho_Z = \frac{3Z}{2\pi r_Z^3} = 2 \times 10^{-9} \text{ fm}^{-3}. \tag{4.150}$$

The corresponding energy density (obtained by multiplying by the nucleon mass) is

$$\epsilon_Z = 3.4 \times 10^6 \text{ gm/cm}^3. \tag{4.151}$$

The corresponding lattice energy is

$$E_{\text{Lat}} = -\frac{9}{10} \frac{(Ze)^2}{r_Z} = 0.1 \text{ MeV}. \tag{4.152}$$

This is large compared to the temperature, either after a month and certainly after a million years. So, fully ionized iron nuclei can exist in a Coulomb lattice at densities in the crust even as low as $\rho_Z$, which is five orders of magnitude lower than the drip density. Down to this density, and possibly lower, the cold approximation is still good.

Nearer the surface of the star, where atoms are not crushed, we have to consider the ionization energy. The binding of an electron in the lowest Bohr orbit in iron is

$$B = \frac{Ze^2}{2r_Z} = 1 \times 10^{-2} \text{ MeV}. \tag{4.153}$$

The total ionization energy must be less than $Z = 26$ times this or less than 0.3 MeV. This statement, however, is too vague to proceed. We use a well known approximation to the Thomas–Fermi binding energy of electrons in an atom [98, 99]. It is

$$B = 1.4 \times 10^{-5} Z^{2.39} = 3.4 \times 10^{-2} \text{ MeV}. \qquad (4.154)$$

This is not small compared to the temperature after a million years so atoms will not be fully ionized by heat at the surface. Will they form a Coulomb lattice as they did at higher densities? Is the lattice energy of partially thermally ionized iron atoms sufficiently large compared to surface temperature to sustain a lattice?

The above questions cannot be answered by the means we have used. Both temperature and density (pressure) will influence the degree of ionization and the solidity or melting of the lattice. The lattice energy will be much smaller than the values computed above for the inner crust because the cell size, $R$, will be larger and the charge, $Z'$, on the ion smaller than $Z$. Lattice melting is related to the ratio of lattice to thermal energy $\Gamma \equiv E_{\text{lat}}/T$.[19] However, it is not really known what value of $\Gamma$ would correspond to the melting point.

Temperature and pressure are so much greater on the surface and into the crust of a compact star than are attainable in the laboratory. So is the degree of ionization for atoms under immense pressure as compared to free atoms. The problem is one of atomic structure under conditions not attainable by laboratory experiment. We do not attempt such an estimate here. Instead we note that in a cold star with mass $1.5M_\odot$ and radius 10 km, the distance from the surface to the radius at which cold matter is fully ionized (4.151) is 76 cm. We conclude that the molten surface of a hot star is very thin and consists of fully ionized atoms.

## 4.12    Stellar Sequences from White Dwarfs to Neutron Stars

An approximate overall picture of the relative densities of white dwarfs and neutron stars is shown in Fig. 4.9, where the mass as a function of the central density is plotted. The BPS equation of state is used for white dwarfs up to the central density just above normal nuclear density marked as $\epsilon_0$. The equations of state for He, C, or O white dwarfs or any mixture of these elements would each yield a somewhat different sequence with maximum masses as high as $1.4M_\odot$. As explained earlier, white dwarfs do not correspond to a single sequence because of the individuality conferred on them by the particular evolution of their progenitors (page 103).

White dwarfs become unstable above the Chandrasekhar mass because the chemical potential of the increasingly relativistic electrons rises along

---

[19]$\Gamma$ is estimated as $\sim 170$ MeV in Ref. [100].

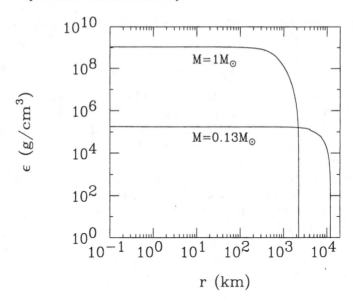

FIGURE 4.11. Mass–energy distribution in two white dwarfs of the BPS equation of state. [101]

the sequence of increasing central density to the point where electron capture by protons (inverse beta decay), called neutronization, yields an energetically more favorable composition. Loss of electron pressure terminates the stable sequence of white dwarfs. This occurs at a central density of about $10^9$ gm/cm$^3$ in the case of the BPS equation of state and about $6 \times 10^{10}$ gm/cm$^3$ for C dwarfs.

Stability is not regained along the sequence until, with increasing density, the pressure of degenerate neutrons provides sufficient support against gravity, initiating the stable neutron star sequence. This sequence begins at densities slightly lower than the density of symmetric nuclear matter.

The distributions of mass–energy in two white dwarfs of the BPS sequence are shown in Fig. 4.11; one is near the mass limit of the BPS equation of state, the other a very light dwarf. The decrease of radius of the star with increasing mass—characteristic of degenerate gravitationally bound objects—is apparent, as is the very flat energy profile in the stellar interior.

Two extreme neutron star sequences are shown in Fig. 4.9. The sequence with maximum mass $\sim 3.2 M_\odot$ corresponds to a choice of equation of state that maximizes the limiting mass subject only to very general constraints on the equation of state—the causal limit equation of state. It will be discussed in connection with an upper bound on neutron star masses (Section 4.18); it is a very stiff equation of state. The other sequence, with lower limiting mass, corresponds to the soft equation of state developed

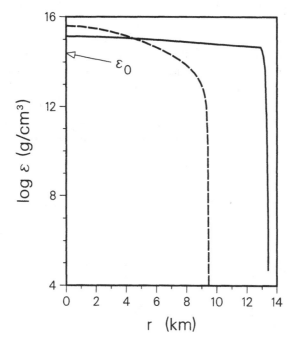

FIGURE 4.12. Mass–energy distribution in the maximum mass neutron stars of the two sequences of Fig. 4.9. The solid curve corresponds to $M \approx 3.2 M_\odot$ and the dashed to $M \approx 0.7 M_\odot$, bracketing realistic values, because the solid curve corresponds to the stiffest possible equation of state consistent with causality, and the dashed, curve to a very soft equation of state, a charge neutral ideal gas of n,p,e.

earlier in this chapter, an ideal gas of neutrons, protons, and electrons that is charge-neutral and in beta equilibrium. Because the Fermi pressure, in this case, is the only agent of resistance to compression, a *lower bound* on the maximum mass limit of neutron stars is $\sim 0.7 M_\odot$ . Nuclear forces with their short-range repulsion will stiffen the equation of state and increase the limiting mass.

Nuclei are bound by the attraction of the nuclear force and have a saturation density that is nearly the same in all nuclei because of the short-range repulsion. In contrast, neutron stars are bound by gravity, and the density is so high that the average nucleon experiences a net repulsion from its interaction with others. Indeed the energy required to compress nucleons to the density found in the cores of neutron stars is 200–300 MeV per nucleon; the nuclear force reduces the binding of neutron stars. In ref [4], we will study sequences of neutron stars corresponding to equations of state which respect certain key nuclear properties, whose values are not strictly determined.

The distribution of mass–energy in the limiting–mass neutron stars of these two extreme sequences are shown in Fig. 4.12. The ideal gas stars near the limiting mass have central densities higher by almost a factor of three than those of the stiff equation of state. They also are much more compacted because of their soft equation of state. The profiles of realistic models generally fall between these extremes. In any case, the envelope of the star, comprising densities less than nuclear, is thin—less than 2 km for the star made of matter with limiting softness, and a fraction of a kilometer for the star of limiting stiffness. Lighter stars in each sequence, those with lower central densities, have larger radii and the envelope is thicker (cf. Fig. 4.9).

## 4.13    Density Distribution in Neutron Stars

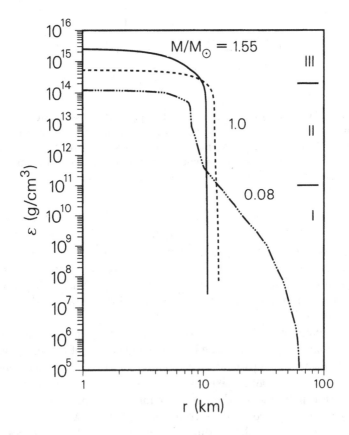

FIGURE 4.13. The distribution of mass density in three neutron stars. Region I corresponds to densities up to the neutron drip region ($\epsilon_{drip} = 4 \times 10^{11}$ g/cm$^3$, II from there to nuclear density $\epsilon_0 = 2.51 \times 10^{14}$ g/cm$^3$, and III above nuclear density.

The distribution of matter density in neutron stars of three different masses are shown in Figure 4.13. The lightest stars has a central density below the saturation density of nuclear matter and is very extended in radius. Those at the maximum mass are very compact. However, stars in the more typical mass range have sharp edges and rather flat density through the interior. This is a reflection of the stiffness, or resistance to compression, of dense matter. Nowhere else in nature is such an exceedingly broad range of densities encompassed in a single object.

## 4.14   Baryon Number of a Star

It is particularly obvious how theorem (3.110) can be applied to a static star to find its baryon number and its binding energy compared to the energy of the dispersed baryons. Let $j^\mu$ denote the conserved baryon number current. It is related to the proper baryon number density, $\rho$ in a local inertial frame of a fluid element of the star at $r$ by

$$j^\mu(r) = u^\mu \rho(r) = e^{-\nu(r)} \rho(r) \delta_0^\mu \qquad (4.155)$$

where, as in (3.163), $u^\mu$ is the fluid four-velocity. In Section 3.3.4 we learned that $\sqrt{-g}\, d^4x$ is the invariant volume element. For the Schwarzschild metric we have

$$\sqrt{-g} = e^{\nu(r)+\lambda(r)}\, r^2\, \sin\theta . \qquad (4.156)$$

Using (3.110) we obtain for the total baryon number of the star

$$
\begin{aligned}
A &= 4\pi \int_0^R e^{\lambda(r)} r^2 \rho(r)\, dr \\
&= 4\pi \int_0^R \left(1 - \frac{2M(r)}{r}\right)^{-1/2} r^2 \rho(r)\, dr .
\end{aligned}
\qquad (4.157)
$$

From whatever theory that provided the equation of state, we know $\rho(r)$ for any $p(r)$ or $\epsilon(r)$. We can therefore integrate the above equation concurrent with the Oppenheimer–Volkoff equations. Thus, the total number of baryons in a star can be computed.

It is also relevant to a description of neutron stars in which not only the neutron and proton are present, but also higher-mass baryons—the hyperons. These higher mass nucleons, some of them containing strange quarks, are likely to be present in the dense interior where the neutron chemical potential may exceed the masses of nucleons, such as the $\Lambda$. This is studied in a later section as well as in detail in reference [4].

The *proper density*[20] of each particle species in the stellar model at a

---

[20]Proper density is used to denote the density in a local inertial frame at the specified point in the gravitational field.

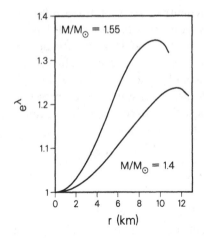

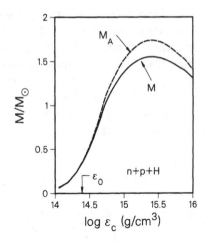

FIGURE 4.14. Metric function interior to neutron stars of two masses. The larger is at the mass limit. See reference [4] for details. ($K$=240 MeV, $m^* = 0.78m$, $x_\sigma = 0.6$).

FIGURE 4.15. For the same case as Fig. 4.14, the gravitational mass $M$ is compared with the mass of the corresponding number of neutrons dispersed at rest to infinity $M_A$.

point $r$ can be found by reference to the particle composition of the calculated equation of state associated with the energy density or pressure for that point. The total number in the star of any such baryon type is given by a formula analogous to the one above, with $\rho$ replaced by the number density of the baryon of interest. Fig. 4.14 shows an example of the metric function $e^{\lambda(r)} = \sqrt{-g_{11}(r)}$ that appears in the integral 4.157.

## 4.15   Binding Energy of a Neutron Star

Because we are able to compute the baryon number, as above, we can compute the mass of the equivalent numbers of neutrons dispersed at rest at infinity,

$$M_A = Am_n . \tag{4.158}$$

The mass $M_A$ is sometimes called the *baryon mass* of the star. In Fig. 4.15 we compare the gravitational masses with the baryon masses. By reading the mass difference from the graph at the limiting mass star and its baryon number from Fig. 4.2, one can find that the binding per neutron is about 100 MeV. This compares with our classical estimate of 160 MeV from (4.25), which actually is the classical gravitational energy. The latter refers only to the gravitational attraction and does not include the work required to overcome the Fermi pressure and the resistance to compression of high–

density matter. The difference

$$B = M_A - M \, , \qquad (4.159)$$

which is the binding energy of the neutron star, contains all such effects and is the net energy gained by assembling the $A$ nucleons from infinity to form an equilibrium neutron star.

The binding energy is the energy released when the core of an evolved star of 10 or more solar masses collapses. About one percent actually appears as the kinetic energy of the supernova explosion in which the neutron star is formed. Ninety-nine percent is carried off by neutrinos in the first few seconds following birth of the neutron star, and a small fraction of one percent appears as light in the supernova remnant.

The baryon mass is usually less than the gravitational mass and always less for configurations near the limiting mass. However, less massive neutron stars have a larger mass than that of all the individual baryons dispersed to infinity. This fact is shown in Fig. 4.16 where the binding energy per nucleon, $B/A = (M_A - M)/A$, is plotted. These stars have even higher gravitational mass as compared to the mass of the equivalent baryon number disassembled in the form of iron nuclei at infinity. Nevertheless, even in the case in which a star is not bound with respect to a simultaneous removal of all constituents to infinity, it is bound with respect to removal of single nucleons (or groups of nucleons, but not all nucleons). The reason for this is the long-range attraction of all the remaining nucleons in the star. From page 95 the energy of a nucleon in the star is $m(1 - 2M/R)^{1/2} = $ const, which is less than $m$. Thus, Oppenheimer–Volkoff stars will not spontaneously disassemble even if not bound in the sense that $M > M_A$. Such neutron stars are metastable with an essentially infinite barrier for disintegration.

It may be relevant to observe that there may exist no creation mechanism for low-mass neutron stars having small or negative binding, even though they are stable in the sense discussed above. They must evade collapse to a black hole. However, the neutrinos carry almost all the binding energy, and because of their weak interaction with nucleons, transmit very little to the infalling material of the defunct star. Thus a low binding energy may fail to eject the ten or more solar masses of the original star, and there is no chance to do so at all if $B/A$ is negative. A large release of binding energy, only a small part of which can be transmitted to the infalling material because of the weak interaction of neutrinos that carry the energy, is required to eject most of the progenitor star and evade the formation of a black hole. This observation may account for the apparent absence of low-mass neutron stars (say less than $1 M_\odot$). It is also likely that stellar cores do not collapse until a Chandrasekhar mass of degenerate material is evolved.

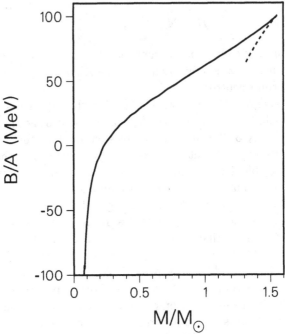

FIGURE 4.16. Binding energy per baryon of a neutron star sequence. The dashed line represents the unstable configurations with central densities higher than that of the limiting mass star. Note the negative binding for stars with $M < 0.2 M_\odot$. Calculational details are in reference [4]. Same case as in Fig. 4.14.

## 4.16  Star of Uniform Density

In general the equations of stellar structure have to be integrated numerically for realistic equations of state. However there is one analytic case of particular interest because it provides a limiting value of $M/R$ that coincides with the absolute limit for any static star satisfying the Oppenheimer–Volkoff equations. A rigorous analysis yields the result

$$\frac{2M}{R} < \frac{8}{9}, \tag{4.160}$$

independent of the equation of state. It depends only on the structure of the relativistic equations for hydrostatic equilibrium and was obtained by Buchdahl [102, 32].

Not surprisingly, the special case that yields the same limit is a star of uniform energy density. The matter of such a star is obviously hypothetical and unphysical because it must be incompressible to maintain a constant density. On the other hand, the idealization is interesting because the energy distribution in neutron stars near their maximum mass limit is rather uniform, though certainly not constant, as we shall see later when treating realistic equations of state. But most importantly, the idealization illustrates that, even in the hypothetical case of incompressible matter, there is a limit on the maximum possible mass of a stable, compact, relativistic configuration. This interesting and instructive idealization was proposed and solved by K. Schwarzschild (see ref. [103]).

For a star of uniform energy density, say $\epsilon_0$, we have from (3.167)

$$M(r) = \frac{4\pi}{3}\epsilon_0 r^3,$$

$$M = \frac{4\pi}{3}\epsilon_0 R^3. \tag{4.161}$$

The equation for hydrostatic equilibrium (3.180) now reads

$$\frac{dp}{dr} = -\frac{4\pi r}{3}\frac{(p(r)+\epsilon_0)(3p(r)+\epsilon_0)}{1-8\pi\epsilon_0 r/3}. \tag{4.162}$$

Although the energy density (by the hypothesis of incompressibility) is constant, the pressure varies throughout the star, and the central value is larger, the larger the stellar radius (and therefore the larger the mass of the object). Rearrange the above equation as

$$-\int_r^R \frac{dp}{(p+\epsilon_0)(3p+\epsilon_0)} = \frac{4\pi}{3}\int_r^R \frac{r\,dr}{1-8\pi\epsilon_0 r^2/3}. \tag{4.163}$$

Integrate and remember that the edge of the star is defined by $p(R) = 0$. We easily find

$$\frac{p(r)+\epsilon_0}{3p(r)+\epsilon_0} = \sqrt{\frac{1-2M/R}{1-2Mr^2/R^3}}. \tag{4.164}$$

Solve for the pressure,

$$p(r) = \epsilon_0\left[\frac{\sqrt{1-2M/R}-\sqrt{1-2Mr^2/R^3}}{\sqrt{1-2Mr^2/R^3}-3\sqrt{1-2M/R}}\right]. \tag{4.165}$$

The central pressure of the star $p_c$ can be obtained now by setting $r = 0$. Solve the resulting equation (4.164) for $2M/R$:

$$\frac{2M}{R} = 1 - \left(\frac{p_c+\epsilon_0}{3p_c+\epsilon_0}\right)^2. \tag{4.166}$$

The larger the radius of a uniform density star, the more massive it is, and the higher the central pressure must be to support this mass. Take the limit of large $p_c$, but short of infinity. We find

$$\frac{2M}{R} < 1 - \left(\frac{1}{3}\right)^2 = \frac{8}{9}, \tag{4.167}$$

independent of $\epsilon_0$. As we noted previously, the above relationship gives the absolute limit of $M/R$ for any static star.

We can easily compute the corresponding radius and mass for the uniform density model. From (4.161) we obtain

$$\frac{4\pi}{3}\epsilon_0 = \frac{M}{R^3} = \frac{M}{R}\frac{1}{R^2}, \tag{4.168}$$

from which there follows

$$
\begin{aligned}
R_{\text{lim}} &= \left(3\pi\epsilon_0\right)^{-1/2}, \\
M_{\text{lim}} &= \frac{4}{9}R_{\text{lim}}.
\end{aligned}
\tag{4.169}
$$

The above is the limiting mass for stars made of incompressible matter whose energy density is $\epsilon_0$. Others in the sequence have smaller mass, radius, and lower central pressure. The existence of such a limit is intrinsic to the General Theory of Relativity, as discussed in Section 3.6.6, and is absent from Newtonian theory.

It is also of interest to note a more stringent condition on $2M/R$. If matter is such that $p < \epsilon$ (believed to be the case), then (4.166) yields

$$\frac{2M}{R} < \frac{3}{4}. \tag{4.170}$$

The opposite situation $p > \epsilon$ is referred to as ultrabaric matter. Because the energy density of baryonic matter at low pressure is dominated by the rest mass of the baryons, we know that any realistic equation of state must be in the domain $p < \epsilon$ at low pressure (at zero pressure there remains the mass of the baryons). The causality condition that a signal cannot propagate faster that the speed of light requires of the equation of state that $dp/d\epsilon < 1$ [104]. If an equation of state satisfies this condition, it cannot cross into the ultrabaric region. However, whereas the 8/9 limit can be shown to depend only on the structure of the Oppenheimer–Volkoff equations and the finiteness of the pressure, the above limit of 3/4 is derived explicitly for a uniform density star. It is therefore not as secure as the 8/9 limit. However, it lies very close to a limit of 0.78 obtained by Bondi [105] under the reasonable assumptions concerning the equation of state: $\epsilon > 0$, $p > 0$, and $dp/d\epsilon < 1$.

## 4.17   Scaling Solution of the OV Equations

For certain simple equations of state which are of interest to us in several different contexts, the Oppenheimer–Volkoff equations have scaling solutions [8, 106]. To explain what is meant by this, consider the equation of state

$$p = s(\epsilon - a), \tag{4.171}$$

where $s$ and $a$ are constants. This equation of state obviously describes matter that is bound at the energy density $a$ because the pressure is zero at that point. Now note that the Oppenheimer–Volkoff equations

$$\frac{dp}{dr} = -\frac{[p(r) + \epsilon(r)][M(r) + 4\pi r^3 p(r)]}{r[r - 2M(r)]},$$
$$dM(r) = 4\pi r^2 \epsilon(r)\, dr,$$

(4.172)

have the same form in the scaled variables,

$$\bar{p} = p/a, \quad \bar{\epsilon} = \epsilon/a, \quad \bar{r} = \sqrt{a}\, r, \quad \overline{M} = \sqrt{a}\, M.$$

(4.173)

Therefore, if the above equations are solved for the sequence of stars corresponding to the equation of state with some particular value of $a$ (and $s$), then the sequence for any other value of the parameter $a$ can be found (for the same $s$) simply by scaling:

$$R(a') = \sqrt{\frac{a}{a'}}\, R(a), \quad M(a') = \sqrt{\frac{a}{a'}}\, M(a),$$
$$\epsilon_c(a') = \frac{a'}{a}\, \epsilon_c(a),$$

(4.174)

where, as usual, $\epsilon_c$ denotes the central density of the star. Note that if one wants sequences with different $s = dp/d\epsilon$, one must solve the Oppenheimer–Volkoff equations for the desired $s$, and all other sequences with different $a$ can be found immediately by scaling.

We shall be interested in rotating stars. The Kepler frequency is of interest in this connection. The Kepler frequency expresses the balance of centrifugal and gravitational forces on a particle on the equatorial plane at the surface of the star. The condition in classical physics is simply

$$\frac{Mm}{R^2} = m\Omega^2 R,$$

(4.175)

or

$$\Omega_c = \sqrt{\frac{M}{R^3}} \quad \text{(Newton)},$$

(4.176)

which is the Newtonian expression for the Kepler angular velocity. This equation need not hold in General Relativity, but as it turns out, it holds to very good accuracy if the right side is multiplied by the factor $\zeta = 0.625$ [107, 108]. From the above scaling law we can also write for the Kepler angular velocity

$$\Omega_K(a') = \sqrt{\frac{a'}{a}}\, \Omega_K(a).$$

(4.177)

## 4.18    Bound on Maximum Mass of Neutron Stars

The Rev. John Michell explained to the Royal Society (London) in 1783 that if stars should exist whose light could not reach us (black holes), nonetheless, their presence may be revealed by their gravitational effect on luminous companion stars [109]. This is precisely how one attempts to identify black holes in the few solar mass range. There are a few such candidates [110]–[112]. However, it is necessary to be able to distinguish between a black hole candidate and a neutron star that would also be invisible to us unless its pulsed radiation was directed toward us once each rotation. The invisible member of the Hulse–Taylor neutron star binary is an example of a star whose presence was detected solely because of the Doppler shift it produces in the timing of pulses from its pulsar companion. In this case, the mass of the invisible star can be determined very accurately from the orbital motion of the pulsar and it is very nearly the same as the pulsar's mass. But if it were above the limit possible for a neutron star, it would be a black hole candidate. The problem and the direction of its solution were recognized by Zwicky [113] and Ruffini [114]–[118].

In the General Theory of Relativity a maximum mass star with central density a few times nuclear density exists above which mass there are no stable configurations, no matter what the equation of state. We have seen that even if matter were to become incompressible at a certain density, a maximum mass still would exist (Section 4.16). If a compact star, unseen except by its gravitational effects on a luminous companion star, is detected and its mass is inferred to be greater than the limiting mass of neutron stars, then the alternative appears to be that it is a black hole. To place such an identification on the firmest ground, the theoretical limiting neutron star mass must be established in a manner that does not depend sensitively on any details of the equation of state; for we do not know it with confidence above nuclear density.

Rhoades and Ruffini attempted to obtain a limit free of any but very general and accepted assumptions. They were: (1) The equations of stellar structure that follow from General Relativity are valid. (2) Matter is stable against local spontaneous expansion or contraction from equilibrium, (Le Chatelier's principle). The principle can be expressed in terms of the equation of state as $dp/d\epsilon \geq 0$ (the pressure must rise as the density or at least remain constant as in the mixed phase of a first-order phase transition in a single-component substance). (3) Disturbances propagate with speed $\sqrt{dp/d\epsilon} < 1$ (referred to as the causal constraint). (4) The high–density equation of state satisfying the above constraints matches continuously to a low–density equation of state that ultimately describes ordinary matter.

For the maximum mass star, the energy profile, as a function of distance from the center, falls very rapidly to zero at the edge of the star, and very little mass is contained in the thin envelope of low–density matter. So the dependence on a specific choice for the low–density domain is weak.

It is fairly obvious (and will become abundantly clear in our later studies) that the stiffer the equation of state (meaning the more rapidly the pressure rises as a function of density) the greater the limiting mass. A high–density equation of state that is at its causal limit will support a greater limiting mass than softer ones. Any density range of matter $\epsilon_1 < \epsilon < \epsilon_2$ for which $dp/d\epsilon$ vanishes is excluded from a star because the Oppenheimer Volkoff equations inform us that the pressure is a monotonically decreasing function of the distance from the center. This is a well-known property of the atmosphere of the earth. The pressure of matter at any radius must support the matter above it against the gravitational attraction of all that lies within. Therefore a constant pressure region in the equation of state contributes nothing to stellar masses. The matter described by such a region, usually a mixed phase, is totally absent from a star. The equation of state at higher density $\epsilon > \epsilon_2$ beyond such a constant pressure region is obviously softer than one that increases with density at $\epsilon_1$. So for the purpose of obtaining an upper bound on the maximum mass, we may consider only equations of state that increase monotonically with density.

Accordingly, let us adopt the BPS equation of state [86, 87, 88] for the low–density domain and match it continuously in density and pressure to an equation of state that is at the causal limit. Let the matching point be denoted by a subscript "$f$". Then the energy density and pressure given by

$$\epsilon(\rho) = \tfrac{1}{2}\Big[\epsilon_f - p_f + (\epsilon_f + p_f)\Big(\frac{\rho}{\rho_f}\Big)^2\Big]$$
$$p(\rho) = \epsilon(\rho) - \epsilon_f + p_f, \qquad \rho \geq \rho_f, \tag{4.178}$$

obey the thermodynamic relationship $p = \rho^2 \partial(\epsilon/\rho)/\partial\rho$. This equation of state is obviously at the causal limit $dp/d\epsilon = 1$, and it is equal to $(\epsilon_f, p_f)$ at $\rho_f$.

There remains the choice of matching density below which a specific equation of state is chosen, in this case BPS. There will be some sensitivity to this choice, and we need to explore it. Let us write the above causal limit equation of state as

$$p = \epsilon - a, \qquad a \equiv \epsilon_f - p_f. \tag{4.179}$$

Now recall that the Oppenheimer–Volkoff equations have scaling solutions for such an equation of state (Section 4.17). If the above causal limit equation of state held for all densities, there would be perfect scaling, that is, when the Oppenheimer–Volkoff equations were solved for a particular $a$ and the mass and radius of any star in the sequence, with central density $\epsilon_c(a)$ determined, the corresponding values for any other $a$ would be given by (4.174).

However, because we use the causal limit equation of state only for $\rho > \rho_f$, there is a thin envelope of matter with $\rho < \rho_f$ overlying the major part of the star. This envelope does not obey the scaling rule, but it contributes

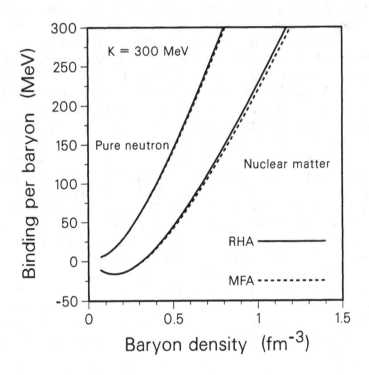

FIGURE 4.17. Binding energy per baryon. The label MFA and RHA are computed with and without vacuum renormalization, with coupling constants chosen to yield the identical nuclear matter properties.

little in mass or radius, and this statement will be more accurate the smaller $\rho_f$. On the other hand, the equation of state is certainly not at the causal limit at densities as low as the saturation density of nuclear matter $\rho_0 = 0.153$ fm$^{-3}$ or, equivalently, $\epsilon_0 \approx 2.5 \ 10^{14}$gm/cm$^3$ because the energy per nucleon has a minimum at saturation density in symmetric matter, and even pure neutron matter reflects attraction near saturation density (cf. Fig. 4.17). So the matching density ought to be chosen larger than saturation density $\rho_0$.

By numerical integration of the Oppenheimer–Volkoff equations using as a matching density one of the grid points in the tabulation of BPS, $\rho_f = 0.2715$ fm$^{-3}$ (which corresponds to $\epsilon_f = 4.636 \ 10^{14}$ gm/cm$^3$ $= 3.442 \ 10^{-4}$ km$^{-2}$ and $p_f = 6.103 \ 10^{33}$ dyne/cm$^2$ $= 5.041 \ 10^{-6}$ km$^{-2}$), we find for the *limiting* mass star

$$M = 3.14 M_\odot, \qquad R = 13.4 \text{ km} \qquad \text{(causal limit)}. \tag{4.180}$$

From the above numbers it is clear that in the vicinity of saturation density the energy density dominates over the pressure by two orders of magnitude so that to good approximation $a$ is just the energy density at the matching point. More accurately,

$$a = 3.392 \ 10^{-4} \text{ km}^{-2} = 4.569 \ 10^{14} \text{ g/cm}^3 \,.$$

So,

$$R(a') = 13.4 \sqrt{\frac{4.569 \ 10^{14} \text{ g/cm}^3}{a'}} \text{ km} \,,$$

$$M(a') = 3.14 \sqrt{\frac{4.569 \ 10^{14} \text{ g/cm}^3}{a'}} \ M_\odot \,. \tag{4.181}$$

If we were to choose the saturation density as matching density (as remarked above it would be hard to justify such a small value), we would obtain $4.3 M_\odot$ for the theoretical upper bound of a neutron star mass. It is safe to conclude that neutron stars cannot have masses that exceed this value and probably cannot exceed the first estimate of $3.14 M_\odot$. With the addition of any other constraints in the form of a realistic equation of state, the bound can only become smaller.

We conclude that the Rhoades–Ruffini bound on the maximum possible mass of a neutron star is a good one. Any compact object of mass greater than $\sim 3 M_\odot$ cannot be a neutron star unless it is rotating at or very near its Kepler frequency (the frequency at which the centrifugal force is balanced by the gravitational force at the equator); in that case the mass limit could be increased by 10 to 20% [119, 120].

For the purpose of identifying black hole candidates, it should be emphasized that the estimate of an upper bound on the mass limit must be conservative. On the other hand, we are also interested in neutron stars themselves in which case we should incorporate as much knowledge or physically motivated theoretical consideration as possible. These can only lower the bound. So actual neutron stars are likely to have masses that are less than the bound derived above, perhaps considerably less. We study this question in the context of realistic constraints on the equation of state in reference [4].

One final note to this discussion. The line $p = \epsilon$ is often referred to as the causal limit, particularly in figures showing equations of state in the $p - \epsilon$ plane. This terminology suggests that equations of state below this line are all causal. This is not so. In (4.179) we see an example that lies below $p = \epsilon$ but is itself at the causal limit $dp/d\epsilon = 1$. What is true is that an equation of state that is at any point below $p = \epsilon$ cannot cross into the region above without having slope greater than unity, becoming therefore acausal. For these reasons, the region above $p = \epsilon$ is better referred to as ultrabaric, meaning simply $p > \epsilon$. Because cold, low–density matter, say iron, has mass but no internal pressure under normal conditions, it lies below the ultrabaric line, and, because causal equations of state (in the sense taken here) cannot cross this line, it is likely that $p < \epsilon$ holds in general.

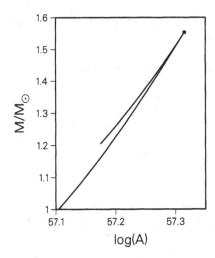

FIGURE 4.18. Mass of neutron stars as a function of baryon number A. Central density increases from low A up to the dot, which is the limiting mass of the stable neutron stars. Central density further increases along the upper curve toward lower values of A. The upper curve corresponds to stars with central densities beyond that of the star at the mass limit. The lower curve represents stable stars.

## 4.19  Stability

Solutions of the Oppenheimer–Volkoff equations correspond to stellar configurations that are in hydrostatic equilibrium. That is clear from the derivation of the equations, as was pointed out in Section 4.5. Equilibrium does not assure stability however. Equilibrium configurations may correspond either to a maximum or to a minimum in the energy with respect to radial compression or dilation. Figure 4.18 illustrates that for a certain range in baryon number of the star, there are at least two equilibrium solutions, one at higher energy and central density than the other. In this section we discuss how to determine whether an *equilibrium* configuration is actually *stable*.

A continuum of equilibrium configurations is shown in Fig. 4.6 which correspond to an equation of state that is smooth and satisfies the condition of *microscopic* stability of matter $dp/d\rho > 0$ (Le Chatelier's principle in its strong form, i.e., without the equality). Two regions along the continuum of compact stars are stable [43]. These are the white dwarf and neutron star regions. There is a broad range extending over many orders of magnitude in central density and in stellar sizes between these two families for which there are no stable stars.

Stars in the unstable region between white dwarfs and neutron stars are subject to vibrational modes that will either cause their disassembly or else their collapse to neutron stars. Indeed, some neutron stars are believed to have been formed by accretion-induced collapse of a white dwarf. At densities below the white dwarf region, there is a long region of instability for cold objects until one reaches other stable bodies, such as brown dwarfs and planets. However, above the stable neutron star sequence, there are no additional regions of stability if the equation of state is of polytropic form, or more generally if it is sufficiently smooth. The physical reasons for this are discussed in Section 4.20. We discuss the necessary and sufficient condition for stability in the next two sections.

## 4.19.1   NECESSARY CONDITION FOR STABILITY

We will prove that along the sequence of equilibrium configurations of the Oppenheimer–Volkoff equations, perfect fluid stars can pass from stability to instability with respect to any radial mode of oscillation only at a value of the central density at which the equilibrium mass is stationary,

$$\frac{\partial M(\epsilon_c)}{\partial \epsilon_c} = 0. \tag{4.182}$$

This is a key result in the discussion of stability. Moreover, the baryon number of an equilibrium star is stationary at the same central densities as the mass [43]. The proof of the last statement follows from (4.47), which we write as

$$\frac{\partial M}{\partial A} = \mu_{\text{Fe}} \left( 1 - \frac{2M}{R} \right)^{1/2}. \tag{4.183}$$

This can be rewritten as

$$\mu_{\text{Fe}} \left( 1 - \frac{2M}{R} \right)^{1/2} \frac{\partial A}{\partial \epsilon_c} = \frac{\partial M}{\partial \epsilon_c}, \tag{4.184}$$

which proves the point.

We now give an heuristic proof of (4.182) which at the same time will provide a *necessary* but not sufficient condition for stability of stars along the sequence of equilibrium configurations.

Refer to Fig. 4.19 for a depiction of a sequence of stars in the vicinity of a maximum. Suppose an equilibrium solution S is perturbed so that the central density $\epsilon_c$ is increased; the star is displaced to C (compressed). The equilibrium star with this new central density lies in mass at $C^\star$. The star at C therefore has a deficit of mass with respect to the equilibrium configuration, a deficit such that gravity under balances the increased central pressure corresponding to the increased density at C, as compared to the equilibrium configuration of the star of this mass that lies at S. The gravitational force acting on the matter of the star will therefore act to return it toward S. A similar argument shows that, if the star at S undergoes a perturbation that decreases its central density, the force acting on the matter of the star will act to return it to S.

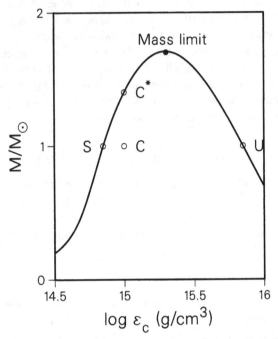

FIGURE 4.19. Schematic of solution of the Oppenheimer–Volkoff equations in a region of increasing and decreasing mass as function of central density. Hydrostatically stable configurations have positive slope so the stable region ends at "Mass limit". See the discussion in the text.

Going through a similar analysis of the star U on the decreasing curve of $M$ vs. $\epsilon_c$, we arrive straightforwardly at the conclusion that, if U is either compressed or decompressed by a radial perturbation, the force acting on the perturbed star will act to drive it further from U, that is to say, equilibrium configurations must satisfy

$$\frac{\partial M(\epsilon_c)}{\partial \epsilon_c} > 0 \quad \text{(necessary condition for stability).} \qquad (4.185)$$

This is a necessary but not sufficient condition for stability.

Thus we have proven that the passage from stability to instability along a sequence of equilibrium configurations occur only at the turning points (4.182). Incidentally, Fig. 4.18 shows that the number of baryons in a star, $A$, reaches its turning point at the maximum mass star, as proven above, and that stars with central density higher than that of the limiting mass star (marked by the dashed line), have higher mass than those of the same $A$ on the lower branch. We shall refer to the results proven above—the coincidence of the turning point of both mass and baryon number and the change of stability of a mode of oscillation at each turning point—as the "turning points theorem".

We now apply the necessary condition for stability to polytropes. Recall that the mass given by (4.118) is an increasing function of $\epsilon_c$ for $\gamma > 4/3$, is stationary at $\gamma = 4/3$, and is a decreasing function for $\gamma < 4/3$. So a necessary condition for the stability of polytropes is

$$\gamma > 4/3. \tag{4.186}$$

The maximum in the mass is known as the mass limit, or limiting mass. White dwarf and neutron star sequences both have a limiting mass, but for different reasons that were discussed in Section 4.9.7.

## 4.19.2   NORMAL MODES OF VIBRATION: SUFFICIENT CONDITION FOR STABILITY

The only astronomically interesting objects are those that are not only in hydrostatic equilibrium, but are also stable against radial oscillations. Stars can vibrate, or ring, like most objects if perturbed. The vibration may either grow or be damped exponentially. To determine which, we solve for the eigenfrequencies of the normal radial modes of vibration. (Any other vibrations can be analyzed in terms of the normal modes.) We refer to the various methods described by Bardeen, Thorne, and Meltzer [121].

For the metric of a spherically symmetric star (3.137), the adiabatic motion of the star is expressed in terms of an amplitude $u_n(r)$ of normal modes of vibration by

$$\delta r(r, t) = e^\nu u_n(r) e^{i\omega_n t} / r^2 \tag{4.187}$$

which denotes small perturbations in $r$. Here $n$ is a mode index with $n = 0$ being the fundamental or nodeless mode. The quantities $\omega_n$ are the star's oscillatory eigenfrequencies. We want to find these eigen-frequencies for each equilibrium configuration of the stellar sequence under discussion.

The eigenequation for $u_n(r)$ which governs the $n$th normal mode—first derived by Chandrasekhar [91]—has the Sturm–Liouville form

$$\frac{d}{dr}\left(\Pi \frac{du_n}{dr}\right) + (Q + \omega_n^2 W)\, u_n = 0, \tag{4.188}$$

where the functions

$$
\begin{aligned}
\Pi &= e^{(\lambda+3\nu)} r^{-2} \Gamma p\,, \\
Q &= -4 e^{(\lambda+3\nu)} r^{-3} \frac{dp}{dr} - 8\pi e^{3(\lambda+\nu)} r^{-2} p\,(\epsilon + p)\,, \\
&\quad + e^{(\lambda+3\nu)} r^{-2} (\epsilon + p)^{-1} \left(\frac{dp}{dr}\right)^2\,, \\
W &= e^{(3\lambda+\nu)} r^{-2} (\epsilon + p)\,, \\
\Gamma &= \frac{d\ln p}{d\ln\rho} = \frac{(\epsilon + p)}{p}\frac{dp}{d\epsilon} \tag{4.189}
\end{aligned}
$$

are expressed in terms of the equilibrium configuration of the star. [Note that the last equality follows from (4.78)]. The quantities $\epsilon$, $p$, and $dp/dr$ denote the energy density, the pressure, and pressure gradient as measured by a local observer. They are obtained from the Oppenheimer–Volkoff equations. The symbol $\Gamma$ denotes the varying adiabatic index at constant entropy. The boundary conditions are $u_n \sim r^3$ at the star's origin and $du_n/dr = 0$ at the star's surface. The second of these assures that the Lagrangian change[21] in the pressure at the surface is zero. Solving the eigenvalue equation leads to the frequency spectrum, $\omega_n^2$ ($n = 0, 1, 2, \ldots$) of the normal radial modes.

A characteristic feature of the Sturm-Liouville equation is that the squared eigenfrequencies $\omega_n^2$ form an infinite discrete sequence, $\omega_0^2 < \omega_1^2 < \omega_2^2 < \cdots$. If any of these is negative for a particular star, the frequency is purely imaginary and therefore any perturbation of the star will grow exponentially in amplitude of oscillation as $e^{|\omega|t}$. Such stars are unstable.

As a consequence of the ordering of the eigenfrequencies and the fact that one mode of oscillation changes in stability at every stationary point (4.182) we may infer the following. If the fundamental mode ($n = 0$) becomes unstable at the maximum neutron star mass, and it does, then at the next minimum in the sequence, either stability is restored to the fundamental mode or the next ($n = 1$) mode becomes unstable, and so on.

The changes in density that arise because of the radial vibrations may produce changes in the local particle populations. We have seen how the neutron to proton fraction varies with density in the ideal gas model in earlier sections of this chapter. In a more realistic model of dense matter, additional baryon species will be populated, or changes in the phase of matter may occur at specific densities such as quark deconfinement as discussed later in the chapter and in reference [4]. Such local changes might be thought to prevent, or at least complicate, the determination of the points in a stellar sequence for which a change between stability and instability occurs. However, this is not so. The period of vibration becomes infinitely long ($2\pi/\omega$) at the points where changes in stability occur ($\omega = 0$). Therefore the location of such points does not depend on the reaction rates.

## 4.20    Beyond the Maximum-Mass Neutron Star

Can superdense stars above the density found in neutron stars exist? The general belief is, or has been, that there are no stable configurations beyond the stable neutron stars [43]. The Oppenheimer–Volkoff equations provide the equilibrium configuration at each central density corresponding to an equation of state of cold matter. There are two density ranges for which

---

[21]The Lagrangian perturbation follows moving fluid elements in the perturbed and unperturbed configurations that correspond to one another. In contrast, Eulerian perturbations refer to changes at fixed points in space.

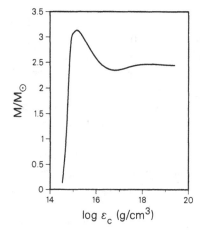

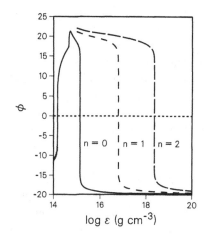

FIGURE 4.20. Stellar sequence corresponding to the causal equation of state of Section 4.18. The first segment with positive slope represent the stable neutron stars. The second is the first in a denumerable infinity of unstable configurations.

FIGURE 4.21. Stability analysis showing a function $\phi$ which has the sign of $\omega^2$ as a function of central density along the sequence of stars of the causal-limit equation of state. (I am indebted to Ch. Kettner for this calculation.)

these configurations are known to be stable. White dwarfs form a stable sequence. They are stabilized by the pressure of degenerate electrons. Neutron stars form a second stable sequence at higher density. They are stabilized by the nucleon Fermi pressure and by the repulsion between nucleons at short range.

Wheeler and collaborators [43] demonstrated that for a *polytropic* equation of state there are a denumerable infinity of stellar sequences that satisfy the necessary condition for stability $dM/d\epsilon_c > 0$, but that beyond the limiting neutron star in density, all configurations are unstable to vibrational modes. Here we show that even the equation of state corresponding to the causal limit of Section 4.18 satisfies the *conditional* theorem of Wheeler et. al.

Figure 4.20 illustrates two segments of the sequence corresponding to the causal-limit equation of state for which the slope is positive. The first segment represents the stable neutron stars. The second is the first in the infinity of segments that satisfy the necessary condition for stability but nonetheless are unstable. That all configurations above the limiting neutron star in density are unstable is shown in Fig. 4.21.[22] The squared frequency

---

[22]The function $\phi$ plotted in Fig. 4.21 has the sign of $\omega^2$ and is defined by $\phi = \mathrm{sgn}(a) \log[1 + |a|]$ where $a \equiv (\omega_n/\mathrm{s}^{-1})^2$.

of the fundamental (nodeless $n = 0$) mode becomes negative at about $10^9$ g/cm$^3$ which terminates the stability of white dwarf configurations. The fundamental mode becomes stable again at $\sim 1.5 \times 10^{14}$ g/cm$^3$, ushering in the stable neutron stars. The same mode becomes unstable at the limiting mass neutron star and remains unstable for all higher central densities. At each succeeding turning point in the mass as a function of central density, whether maximum or minimum, an additional vibrational mode becomes unstable, as illustrated for the $n = 1$ and $n = 2$ modes in Fig. 4.21. This is in accord with the "turning points theorem" stated on page 150.

For those of the sequence shown in Fig. 4.20 with positive slopes that lie higher in central density than the stable neutron star sequence, the energy profile increases so steeply as the center of the star is approached that the star is unstable to radial vibrations. As the amplitude of a radial perturbation grows, the lowest (nodeless) mode may bring the central part of the star within its critical Schwarzschild radius when the amplitude of the vibration is sufficiently high. The core collapses, and then the rest of the star, with nothing to support it, follows. If this does not happen first, then in the outward motion of the vibration, the star may disassemble, or explode. One or the other is the fate of stars more dense than neutron stars if the equation of state is sufficiently smooth, as it might well be.

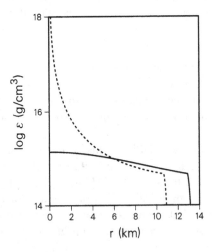

FIGURE 4.22. Mass–energy distribution of the stars at the two maxima of Fig. 4.20. Solid line refers to the first maximum. The star at the second maximum is not really a stable configuration over the long-term; because of the steep rise in density toward the center of the star, a slight perturbation of the star will cause it to collapse gravitationally to a black hole.

A comparison is made in Fig. 4.22 between the mass–energy distribution of a stable neutron star at the Oppenheimer Volkoff limit and that of the unstable star at the next maximum. The remarkably steep energy distribution in the second case is responsible for the instability described above and characterizes the distribution in all stellar configurations above the Oppenheimer–Volkoff limit for neutron stars for most equations of state.

## 4.21 Hyperons and Quarks in Neutron Stars

Although conceived of originally as stars made of neutrons, we have seen that taking account of beta equilibrium, neutron stars are not made purely of neutrons. Neutrons are no doubt the dominant particle species. But as the Fermi energy increases with density, other particle species are populated. At the simplest level, a neutron star could be thought of as an equilibrium admixture of neutrons, protons, and electrons, with an equal number of the latter two so as to preserve charge neutrality. But at higher density additional baryon species will be populated, namely the hyperons, as seen in Figure 4.23, and at sufficiently high density, quarks—which are normally confined in individual baryons and mesons—may be liberated in a larger macroscopic region of colorless quark matter. This topic is covered in detail in reference [4]; here we present some of the salient parts.

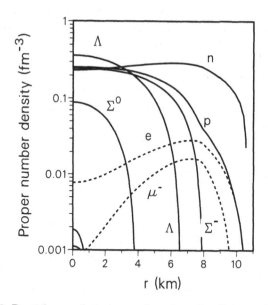

FIGURE 4.23. Particle populations as a function of radial coordinate in a neutron star of mass $1.45 M_\odot$. Note that the neutron is dominant but hyperons form a significant part of the population.

When we considered a star made from neutrons, protons and electrons, we encountered the condition $\mu_e + \mu_p = \mu_n$ for chemical equilibrium in connection with equation 4.83. Two of the chemical potentials are independent, and we may take them to be one for baryon charge number $q_b$, and one for electric charge number $q_e$ (in units of the electron charge). Then each baryon, $B$, has a chemical potential,

$$\mu_B = q_b\mu_n - q_e\mu_e, \tag{4.190}$$

(so that for the proton, $\mu_p = \mu_n - \mu_e$, because the proton has opposite charge to the electron.) For the negative muon,

$$\mu_{\mu^-} = \mu_e. \tag{4.191}$$

Populations of hadrons in a neutron star near the maximum mass (of the particular model) are shown in Figure 4.23. However, depending on the unknown density at which quarks become deconfined, hadrons may becomes dissolved into their quark components. Therefore we add to the above discussion, the chemical potentials for the three quark flavors, which of course can be written in terms of the two independent chemical potentials $\mu_n$ and $\mu_e$, namely

$$\mu_u = (\mu_n - 2\mu_e)/3, \quad \mu_d = \mu_s = (\mu_n + \mu_e)/3. \tag{4.192}$$

The baryon and electric charge content of the quarks is reflected in the signs and factors appearing in the above expressions. The suffixes stand for the quark flavor—up, down, and strange.

Figure 4.24 shows particle number density in a neutron star in which the core of 2 km. is purely quark, the intermediate region extending from 2 to 8 km is in the mixed phase of quarks and hadrons, and the outer two kilometers is purely hadronic, mostly made of neutrons with a small admixture of protons and electrons and negative muons. This figure serves to illustrate the rich particle structure of what we call neutron stars.

## 4.22   First Order Phase Transitions in Stars

Generally, physicists think of a phase transition such as from water to vapor as being *typical* of a first order transition. In the real world it is far from typical. Its characteristics are: if heated at constant pressure, the temperature of water and vapor will rise to 100 C and remain there until all the water has been evaporated before the temperature of the steam rises. This is true of substances having *one* independent component (like pure water). The situation can, and usually is much more interesting for substances with two or more independent components [68]. Such a substance we refer to sometimes as a *complex* substance. Neutron stars are an example. The independent components are the conserved baryon and electric charge. Until about 1990, all authors forced stellar models into the mold of single-component substances by imposing a condition of *local* charge neutrality and ignoring the discontinuity in electron chemical potential at

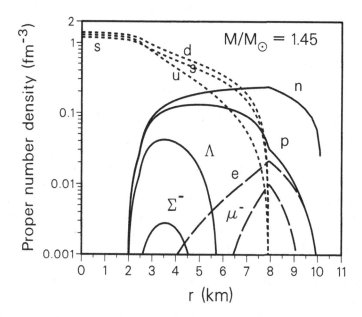

FIGURE 4.24. Particle populations in a neutron star of mass $1.45 M_\odot$ as a function of radial coordinate including quarks. Note that three phases are present, pure quark matter in the center out to 2 km., a mixed quark-hadronic phase out to about 8 km., surrounded by hadronic matter.

the interface of two phases in equilibrium. This procedure cannot satisfy Gibbs criteria for phase equilibrium in complex systems. An altogether new phenomenon is found by solving the problem correctly: Indeed, a *Coulomb crystalline region* involving the two phases in equilibrium could form, an idea that had not previously come to light [68].

## 4.22.1 DEGREES OF FREEDOM AND DRIVING FORCES

Two features can come into play in phase transition of complex substances that are absent in simple substances. The degree(s) of freedom can be seen in the following way. Imagine assembling a star in a pure phase (say ordinary nuclear matter) with $B$ baryons and $Q$ electric charges, either positive, negative or zero. (Of course, more precisely, we consider a typical local inertial region.) The concentration is said to be $c = Q/B$.

Now consider another local region deeper in the star and at higher pressure with the same number of baryons and charges, but with conditions such that the mixed coexistence phase occupies that part of the volume. Suppose the baryons and electric charges in the two phases are distributed such that concentrations in the two phases, labeled by '1' and '2', are

$$c_1 = Q_1/B_1 \quad \text{and} \quad c_2 = Q_2/B_2 \,. \tag{4.193}$$

The conservation laws are still satisfied if

$$Q_1 + Q_2 = Q, \quad B_1 + B_2 = B \,. \tag{4.194}$$

Why might the concentrations in the two phases be different from each other and from the concentration in the other local volumes at different pressure? Because the *degree of freedom* of redistributing the concentration may be exploited by *internal forces* of the substance so as to achieve a lower free energy. In a single-component substance there was no such degree of freedom, and in an $n$-component substance there are $n - 1$ degrees of freedom. In deeper regions of the star, still different concentrations may be favored in the two phases in equilibrium at these higher-pressure locations. So each phase in equilibrium with the other, may have continuously changing properties from one region of the star to another. (This is unlike the simple substance whose properties remain unchanged in the equilibrium phase, until only one phase remains.)

The key recognition is that conserved quantities (or independent components) of a substance are conserved *globally*, but need not be conserved *locally* [68]. Otherwise, Gibbs conditions for phase equilibrium cannot be satisfied. Let us see how this is done.

Gibbs condition for phase equilibrium in the case of two conserved quantities is

$$p_1(\mu_n, \mu_e, T) = p_2(\mu_n, \mu_e, T) \,. \tag{4.195}$$

We have introduced the neutron and electron chemical potentials by which baryon and electric charge conservation are to be enforced. In contrast to the case of a simple substance, for which Gibbs condition—$p_1(\mu, T) = p_2(\mu, T)$—can be solved for $\mu$, the phase equilibrium condition cannot be satisfied for substance of more than one independent component without additional conservation constraints. Clearly, *local* charge conservation $(q(r) \equiv 0)$ must be abandoned in favor of global conservation $(\int q(r)q(r) \equiv 0)$, which is after all what is required by physics. For a uniform distribution global neutrality reads,

$$(1 - \chi)q_1(\mu_n, \mu_e, T) + \chi q_2(\mu_n, \mu_e, T) = 0 \,, \tag{4.196}$$

where

$$\chi = V_2/V, \quad V = V_1 + V_2 \,. \tag{4.197}$$

Given $T$ and $\chi$ we can solve 4.195 and 4.196 for $\mu_n$ *and* $\mu_e$. Thus the solutions are of the form

$$\mu_n = \mu_n(\chi, T), \quad \mu_e = \mu_e(\chi, T) \,. \tag{4.198}$$

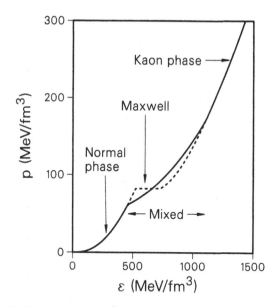

FIGURE 4.25. Solid line: equation of state for neutron star matter with a kaon condensed phase. (The situation is analogous for any phase transition in a complex substance, including the transition from hadronic matter to deconfined quark matter.) Regions of the normal nuclear matter phase, the mixed phase, and the pure kaon condensed phase are marked. Notice that the pressure changes monotonically through the mixed phase. Dashed line: The Maxwell construction with the typical constant pressure region does not satisfy equality of the electron chemical potential in the two phases.

Because of the dependence on $\chi$ $(0 \leq \chi \leq 1)$, all properties of the phases in equilibrium change with proportion, $\chi$, of the phases. This contrasts with simple (one component) substances. These properties are illustrated for the pressure in Figure 4.25. The behavior of the pressure is illustrated for two cases: (1) a simple, and (2) a complex substance. In the latter case, the pressure is monotonic increasing as a function of density or equivalently, position in the star, as proven above. This is in marked contrast to the pressure plateau of the simple (one component) substance. As a consequence of the monotonic behavior of the pressure, the mixed phase occupies a finite radial extent in the star. A constant pressure region, as in the phase transition of a simple substance where the pressure is constant in the transition region, is totally absent from the star, since it occupies a single radial point.

## 4.22.2    ISOSPIN SYMMETRY ENERGY AS A DRIVING FORCE

A well known feature of nuclear systematics is the valley of beta stability which, aside from the Coulomb repulsion, endows nuclei with $N = Z$ the greatest binding among isotones ($N + Z = $ const). Empirically, the form of

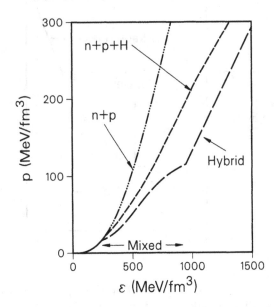

FIGURE 4.26. Equation of state for matter in *beta equilibrium* for three hypothetical models of dense nuclear matter;   (1): only neutrons and protons are present $(n + p)$,   (2): in addition to neutrons and protons, hyperons (H) are also present $(n + p + H)$,   (3): *Hybrid* denotes the equation of state for which matter has a low-density nuclear matter phase (below $\epsilon \approx 200$), an intermediate mixed phase of hadrons and quarks (up to $\epsilon \approx 1000$), and a high-density quark phase (above $\epsilon \approx 1000$). Discontinuities in slope signal the transition between these phases.

the symmetry energy is

$$E_{N-\text{sym}} = \epsilon[(N - Z)/(N + Z)]^2 . \tag{4.199}$$

Physically, this arises in about equal parts from the difference in energies of neutron and proton Fermi energies and the coupling of the $\rho$ meson to the nucleon isospin current. Consider now a neutron star. While containing many nucleon species, neutron star matter is still very isospin asymmetric—it sits high up from the valley floor of beta stability (4.199)—and must do so because of the asymmetry imposed by the large strength of the Coulomb force compared to the gravitational, which enforces charge neutrality on the star.

We may define sample volumes $V$, small enough to be local inertial frames because the relative change in spatial metric is $\sim 10^{-9}$ over a distance of the order of $10^{10}$ internucleon spacings [4]. Therefore all problems of the structure and composition of matter can be solved in Minkowski spacetime and the results used in the matter stress-energy tensor appearing in Einstein's equations. We examine such volumes of matter at ever-deeper depth in a star until we arrive at a volume where the pressure is high enough that some of the quarks have become deconfined; that both phases are present in the local volume. According to what has been said above, the highly unfavorable isospin repulsive asymmetry energy, depending on the isospin $((N - Z)/(N + Z))$ of the nuclear phase of neutron star matter, can be lowered if some neutrons in the hadronic phase exchange one of their d quarks with a u quark in the quark phase that is in equilibrium with it. In this way the nuclear matter will become positively charged and the quark matter will carry a compensating negative charge, and the overall energy will be lowered. The *degree* to which the exchange will take place will vary according to the proportion, $\chi$, of the phases: Clearly a region with a small proportion of quark matter cannot as effectively relieve the isospin asymmetry of a large proportion of neutron star matter of its excess isospin as can a volume of the star where the two phases are in more equal proportion. We see this quantitatively in Figure 4.27 where the charge densities on hadronic and quark matter are shown as a function of proportion of the phases.

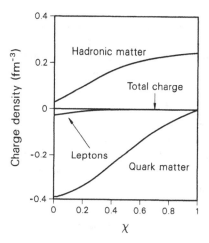

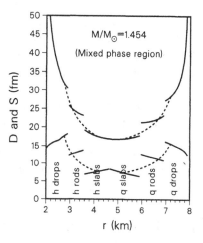

FIGURE 4.27. Electric charge densities on hadronic and quark matter in the mixed co-existence phase, as a function of proportion of one phase compared to the other ($\chi$ volume fraction of quark phase).

FIGURE 4.28. Diameter D (bottom and Spacing S (top) of the geometrical phases as a function of position in the co-existence region of the star.

### 4.22.3  GEOMETRICAL PHASES

In equilibrium, the isospin driving force tends to concentrate positive charge on nuclear matter and compensating negative charge on quark matter. The Coulomb force will tend to break up regions of like charge while the surface interface energy will resist this tendency. The same competition is in play in the crust of the star where ionized atoms sit at lattice sites in an electron sea. For the idealized geometries of spheres, rods, or sheets of the rare phase immersed in the dominant one, and employing the Wigner-Seitz approximation (in which each cell has zero total charge, and does not interact with other cells), closed form solutions exist for the diameter $D$, and spacing $S$ of the Coulomb lattice. The Coulomb and surface energy for drops, rods or slabs ($d = 3, 2, 1$) have the form:

$$\epsilon_C = C_d(\chi)D^2 , \qquad \epsilon_S = S_d(\chi)/D , \qquad (4.200)$$

where $C_d$ and $S_d$ are simple algebraic functions of $\chi$. The sum is minimized by $\epsilon_S = 2\epsilon_C$. Hence, the diameter of the objects at the lattice sites is

$$D = [S_d(\chi)/2C_d(\chi)]^{1/3} , \qquad (4.201)$$

where their spacing is $S = D/\chi^{1/d}$ if the hadronic phase is the background or $S = D/(1 - \chi)^{1/d}$ if the quark phase is background. Figure 4.28 shows the computed diameter and spacing of the various geometric phases of quark and hadronic matter as a function of radial coordinate in a hybrid neutron star.

Figures 4.29 and 4.30 show the radial distribution of the geometric phases and their nature—either quark or hadronic matter—for stars of two different masses, respectively. In the higher mass star, the full complement of geometric phases exists—from a pure quark matter phase through all the geometric phases to pure hadronic matter. By contrast, in the lower mass, only two of the geometric mixed phases are present.

### 4.22.4  COLOR-FLAVOR LOCKED QUARK-MATTER PHASE (CFL)

Alford, Rajagopal and Wilczek have argued that the Fermi surface of the quark deconfined phase is unstable to correlations of quarks of *opposite* momentum and *unlike* flavor and form BCS (Bardeen-Cooper-Schriefer) pairs [122, 123]. Alford et. al. estimate a pairing gap of $\Delta \sim 100$ MeV. The greatest energy benefit is achieved if the Fermi surfaces of all flavors are equal in radius. This links color and flavor by an invariance to simultaneous rotations of color and flavor. The approximate energy density corresponding to the gap is

$$\epsilon_{\Delta-\text{CFL}} \sim C(k_F \Delta)^2 \sim 50 \cdot C \text{ MeV/fm}^3 , \qquad (4.202)$$

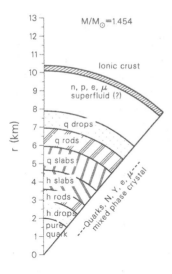

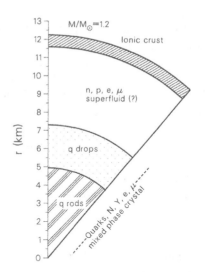

FIGURE 4.29. Pie section showing geometrical phases in star of high mass.

FIGURE 4.30. Pie section showing geometrical phases in star of low mass.

where $C = 3/\pi^2$.

This is another "driving force" as spoken of above *in addition* to the nuclear symmetry energy $E_{N-sym}$. It acts, not to restore isospin symmetry in nuclear matter, but instead to favor color-flavor symmetry in the quark phase. Rajagopal and Wilczek [124] have argued that the CFL phase, which is *identically* charge neutral and has this large pairing gap *may* preempt the possibility of phase equilibrium between confined hadronic matter and the quark phase; that any amount of quark matter would go into the charge neutral CFL phase (with equal numbers of u, d and s quarks, irrespective of mass) and that the mixed phase spoken of above would be absent. That the nuclear symmetry driving force would be overcome by the color-flavor locking of the quark phase leaving the degree of freedom possessed by the two-component system unexploited. The discontinuity of the electron chemical potential in the two phases, hadronic and quark matter would be patched by a spherical interface separating a core of CFL phase in the star from the surrounding hadronic phase. For that conclusion to be true, a rather large surface interface coefficient was chosen by dimensional arguments.

However, my opinion is that nature will make a choice of surface interface properties between hadronic and quark matter such that the degree of freedom of exchanging charge can be exploited by the driving forces (here two in number as discussed below). This is usually the case. Physical systems generally have their free energy lowered when a degree of freedom (as spoken of above) becomes available.

With two possible phases of quark matter, the uniform uncorrelated one discussed first, and the CFL phase as discussed by Rajagopal and Wilczek, there is now a competition between the CFL pairing and the nuclear symmetry-energy densities, and these energy densities are *weighted* by the volume proportion $\chi$ of quark matter in comparison with hadronic matter in locally inertial regions of the star. That is to say, $\epsilon_{CFL}$ and $\epsilon_{sym}$ are not directly in competition, but rather they are weighted by the relevant volume proportions. It is not a question of "either, or" but "one, then the other".

The magnitude of the nuclear symmetry energy density at a typical phase transition density of $\rho \sim 1/\text{fm}^3$ is

$$\epsilon_{N-sym} = 35[(N-Z)/(N+Z)]^2 \text{ MeV/fm}^3 . \tag{4.203}$$

To gain this energy a certain price is exacted from the disturbance of the symmetry of the uniform quark matter phase in equilibrium with it, namely $\epsilon_{Q-sym}$. As can be inferred from Figure 4.27, the price is small compared to the gain. On the other side, the energy gained by the quark matter entering the CFL phase was written above and is offset by the energy not gained by the nuclear matter because the CFL preempts an improvement in its isospin asymmetry. So we need to compare

$$(1-\chi)\epsilon_{N-sym} - \chi\,\epsilon_{Q-sym} - [\epsilon_{surf}(\chi) + \epsilon_{coul}(\chi)] \tag{4.204}$$

with

$$\chi\,\epsilon_{\Delta-CFL} - (1-\chi)\epsilon_{N-sym} . \tag{4.205}$$

The behavior of these two lines as a function of proportion of quark phase $\chi$ in a local volume in the star is as follows:[23] The first expression for the net gain in energy due to the formation of a mixed phase of nuclear and uniform quark matter monotonically decreases from its maximum value at $\chi = 0$ while the second expression, the net energy gain in forming the CFL phase monotonically increases from *zero* at $\chi = 0$. Therefore as a function of $\chi$ or equivalently depth in the star measured from the depth at which the first quarks become deconfined, nuclear symmetry energy is the dominating driving force, while at some value of $\chi$ in the range $0 < \chi < 1$ the CFL pairing becomes the dominating driving force.

In terms of Figure 4.29, several of the outermost geometric phases in which quark matter occupies lattice sites in a background of nuclear matter are undisturbed. But the sequence of geometric phases is terminated before the series is complete, and the inner core is entirely in the CFL phase.

---

[23]The behavior of the quantity in square brackets can be viewed in Figure 9.14 of reference [4],.2'nd ed.

In summary, when the interior density of a neutron star is sufficiently high as to deconfine quarks, a charge neutral color-flavor locked phase with no electrons will form the inner core. This will be surrounded by one or more shells of mixed phase of quark matter in a uniform phase in phase equilibrium with confined hadronic matter, the two arranged in a Coulomb lattice which differs in dimensionality from one shell to another. As seen in Figure 4.27, the density of electrons is very low to essentially vanishing, because overall charge neutrality can be achieved more economically among the conserved baryon charge carrying particles. Finally, all this will be surrounded by uniform charge neutral nuclear matter with varying particle composition according to depth (pressure), (cf. Figure 4.29.)

## 4.23    Signal of Quark Deconfinement in Neutron Stars

Here we develop an signal of quark deconfinement in neutron stars that is, in principle, observable [5, 125]. The deconfined phase of hadronic matter called quark matter is believed to have pervaded the early universe and may reside as a permanent component of neutron stars in their dense high-pressure cores [126, 127, 128, 129, 130, 131]. However no means of detecting its presence has been found because the properties of neutron stars and those with a quark matter core are expected to be very similar. Alternately, instead of looking to the properties of the star itself, we study the spin-down behavior of a rotating star with the realization that changes in internal structure as the star spins down will be reflected in the moment of inertia and hence in the deceleration.

Pulsars are born with an enormous store of angular momentum and rotational energy which they radiate slowly over millions of years by the weak processes of electromagnetic radiation and a wind of electron-positron pairs [132, 133, 134, 135]. When rotating rapidly, a pulsar is centrifugally flattened. The interior density will rise with decreasing angular velocity and may attain the critical density for a phase transition. First at the center and then in an expanding region, matter will be converted from the relatively incompressible nuclear matter phase to the highly compressible quark matter phase. The weight of the overlaying layers of nuclear matter will compress the quark matter core and the entire star will shrink. The mass concentration will be further enhanced by the increasing gravitational attraction of the core on the overlaying nuclear matter. The moment of inertia thus decreases anomalously with decreasing angular velocity as the new phase slowly engulfs a growing fraction of the star.

The decrease of moment of inertia caused by the phase transition is superposed on the natural response of the stellar shape to a decreasing centrifugal force occasioned by radiation loss. Therefore, to conserve angular

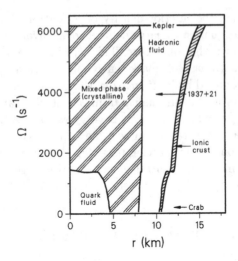

FIGURE 4.31. Evolution of radial location (x-axis) between boundaries of various phases in a neutron star of given baryon number as a function of moment of inertia (y-axis), which is a decreasing function of time in a pulsar that is spinning down. Stellar mass is $1.42 M_\odot$ for the nonrotating counterpart. Formation of a pure quark matter core occurs below $I \sim 107$ km$^3$ ($\Omega \approx 1400$ rad/s). Note the decrease of radius with time. See Ref. [7] for a description of spatial geometry in mixed phase.

momentum not carried off by radiation, the deceleration $\dot{\Omega}$ of the angular velocity must respond by decreasing in absolute magnitude and may actually change sign. The pulsar may spin up for a time, just as an ice skater spins up upon contraction of the arms before air resistance and friction reestablish spin-down.

Such an anomalous decrease of the moment of inertia as described is analogous to the phenomenon of "backbending" in the rotational bands of nuclei predicted by Mottelson and Valatin and observed years ago [136, 137, 138]. The moment of inertia of a nucleus changes anomalously because of a change in phase from a nucleon spin-aligned phase at high angular momentum to a pair-correlated superfluid phase at low. The connection between the moment of inertia and the internal structure of a neutron star is shown in Fig. 4.31 and between moment of inertia and frequency in Fig. 4.32.

The property of asymptotic freedom of quarks makes a phase change from confined to deconfined matter inevitable for sufficiently high energy density of hadronic matter. We have therefore referred to the transition as being from the confined to deconfined phase. However, it is apparent in our discussion and results, that the phenomenon we discuss is not unique to this particular transition. It is simply the most plausible in our view, but any phase transition in dense hadronic matter is interesting.

Now we develop the appropriate measure for detecting the occurrence of

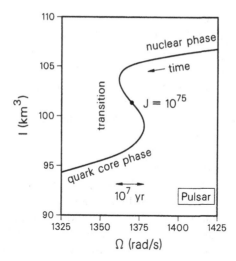

FIGURE 4.32. Moment of inertia as a function of rotational angular velocity over the epoch during which an increasing central volume of the star is passing into the deconfined quark matter phase. The temporal development is from large to small $I$. Spin-up of the pulsar is clearly evident. An analogous phenomenon known as "backbending" is observed in the rotational properties of nuclei.

a phase transition and its envelopment of an increasing proportion of the mass and volume of the star with time. The production of electromagnetic radiation and a wind of relativistic pairs by the rotating magnetic field of the star exert a torque on the star which is characteristic of magnetic dipole radiation. The energy loss equation representing processes of multipolarity $n$ is of the form

$$\frac{dE}{dt} = \frac{d}{dt}\left(\frac{1}{2}I\Omega^2\right) = -C\Omega^{n+1} \qquad (4.206)$$

where $n = 3$ for magnetic dipole radiation. The rate of change of frequency is governed by

$$\dot{\Omega} = -\frac{C}{I(\Omega)}\left[1 + \frac{I'(\Omega)\,\Omega}{2I(\Omega)}\right]^{-1}\Omega^n . \qquad (4.207)$$

For low frequency or if changes in $I$ are ignored this reduces to the usual form quoted in the literature. $\dot{\Omega} = -K\Omega^n$ where $K = C/I$, (cf. Ref. [139, 140]).

The dimensionless measurable quantity $\Omega\ddot{\Omega}/\dot{\Omega}^2$ is referred to as the "braking index". It would be equal to the intrinsic index $n$ of the energy-loss mechanism (4.206) if the frequency were small or if the moment of inertia were a constant. However these conditions are not usually fulfilled and the measurable quantity is not constant. Rather it has the value

$$n(\Omega) \equiv \frac{\Omega\ddot{\Omega}}{\dot{\Omega}^2} = n - \frac{3I'\Omega + I''\Omega^2}{2I + I'\Omega} \qquad (4.208)$$

where $I' \equiv dI/d\Omega$ and $I'' \equiv dI^2/d\Omega^2$. The progression of the new phase through the central region of the star will be signaled by an anomalous value of the braking index, far removed from the canonical value of $n$.

Because the pulsar rotational energy is coupled to weak processes, $\dot{\Omega}$ is small and none of the quantities in (4.208) will change appreciably over any observational time span. However the signal is carried in non-zero derivatives $dI/d\Omega$ and $dI^2/d\Omega^2$; these are large during the phase transition epoch because of the progressive conversion of nuclear matter into compressible quark matter. As can be seen from (4.208), large derivatives of the moment of inertia will produce enormous deviations of the braking index from its canonical value while the region of the star occupied by quark matter is growing. Since the growth is paced by the slow spin-down of the pulsar, we will find that the signal is "on" over an extended epoch.

The behavior of the moment of inertia in the critical frequency interval for our model star (which is described later) is shown in Fig. 4.32. As the pulsar evolves in time (decreasing $I$) the derivative $dI/d\Omega$ passes through two singularities, switching between $+\infty$ and $-\infty$ at each turning point of $\Omega$. From (4.207) it is clear that the deceleration $\dot{\Omega}$ will pass through zero and change sign at both turning points becoming an acceleration in the central part of the 'S'; the pulsar spins up for a time. Moreover, $-I''$ has similar singularities as can be found from

$$-I'' = \left(\frac{dI}{d\Omega}\right)^3 \frac{d^2\Omega}{dI^2} . \qquad (4.209)$$

Consequently $n(\Omega)$ goes to $\pm\infty$ at the two turning points respectively as shown in Fig. 4.33. We have plotted the braking index as a function of $I$ because $I$ decreases monotonically with the time, unlike $\Omega$.

(The spin-up referred to above has nothing to do with the minuscule spin-up known as a pulsar glitch. The relative change in moment of inertia in a glitch episode is very small ($\Delta I/I \sim -\Delta\Omega/\Omega \sim 10^{-6}$ or smaller) and approximates closely a continuous response of the star to changing frequency on any time scale that is large compared to the glitch and recovery interval. Excursions of such a magnitude as quoted would fall within the thickness of the line in Fig. 4.32.)

A change in moment of inertia owing to a change in phase such as we have described is evidently a robust phenomenon. However, we cannot be sure that nature will respond as strongly as our model does. Backbending, as in Fig. 4.32, is an extreme response of the moment of inertia to the progression of a phase transition through the central region of the star. Instead, the transition of the moment of inertia from that of a neutron star (at high $\Omega$) to a hybrid star may be a single-valued function of $\Omega$. However the transition between these two moments can still be marked by a large first derivative $I'$ and a second derivative $I''$ that is large and changes sign in the transition interval. In this case, the braking index will not swing between $\pm\infty$ but nonetheless can attain large positive and negative values.

Typically it is difficult to measure $\ddot{\Omega}$ (and hence the braking index) because of timing noise. Only four braking indices are presently known. However, for a star that is passing through the phase transition epoch, the deceleration (4.207) is reduced markedly (even changing to acceleration). The second derivative must therefore be large in absolute magnitude through all but the central portion of the epoch where spin-up occurs. Consequently, the second derivative of frequency should be easier to measure for pulsars passing through the epoch than for typical pulsars. Hence, difficulty in measuring the second derivative may be used as a de-selection criterion.

We estimate the plausibility of observing in the pulsar population a signal of the kind that we find in our calculation. Assume that neutron stars are created in a narrow mass window (as present evidence suggests). The duration over which the observable index is anomalous is $\Delta T \approx -\Delta\Omega/\dot{\Omega}$ where $\Delta\Omega$ is the frequency interval of the anomaly. The range in which $n$ is smaller than zero or larger than six (Fig. 4.33) is $\Delta\Omega \approx 100$ rad/s. Take a typical period derivative of $\dot{P} \sim 10^{-16}$ s/s to find $\Delta T \sim 10^5$ years. The signal would endure for $1/100$ of a typical active pulsar lifetime. Similarly we estimate that spin-up would last for about $1/6$'th of that time. Given that more than 1000 pulsars are presently known 10 of these may be signaling the growth of a central region of new phase.

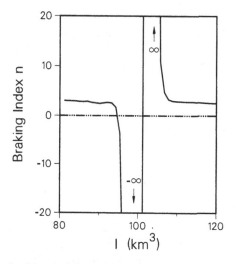

FIGURE 4.33. The braking index as a function of angular velocity signals by its large departures from $n = 3$ the progression of the deconfined phase in the stellar interior. The braking index would be identically 3 if it were not for the response of $I$ to rotation, and in any case tends to 3 for low frequency (for magnetic dipole radiation). The duration of time over which the braking index differs significantly from its canonical value of 3, undergoing a discontinuity, is $10^8$ years. Note the correspondence in moment of inertia $I$ here and in Figure 4.32: The discontinuity of the braking index occurs at the central value of the backbend.

We describe briefly the calculation. The equations describing the configurations of rotating stellar structures are solved for a sequence having the same baryon number but different rotational frequencies. The usual expression for the moment of inertia in General Relativity due to Hartle is not adequate for our purpose (see Ref. [141] equations (49) and (62)). Not only are effects of internal constitutional changes in the star absent but also absent are the effects of the alteration of the metric of spacetime by rotation, the dragging of local inertial frames, and even centrifugal flattening. Instead, to calculate the moment of inertia, we must use an expression that incorporates the above effects as derived by Glendenning and Weber [142, 143].

The stellar model is described as follows. Neutron star matter has a charge neutral composition of hadrons consisting of members of the baryon octet together with leptons when in the purely confined phase. The properties of such matter are calculated in a covariant mean field theory as described in Refs. [85, 144, 145, 146, 147, 148]. The values of the parameters that define the coupling constants of the theory are certain fairly well constrained properties of nuclear matter and hypernuclei as described in the references; (binding energy of symmetric nuclear matter $B/A = -16.3$ MeV, saturation density $\rho = 0.153$ fm$^{-3}$, compression modulus $K = 300$ MeV, symmetry energy coefficient $a_{\text{sym}} = 32.5$ MeV, nucleon effective mass at saturation $m^{\star}_{\text{sat}} = 0.7m$ and ratio of hyperon to nucleon couplings $x_{\sigma} = 0.6$, $x_{\omega} = 0.653 = x_{\rho}$ that yield, together with the foregoing parameters, the correct $\Lambda$ binding in nuclear matter [144]). Quark matter is treated in a version of the MIT bag model with the three light flavor quarks ($m_u = m_d = 0$, $m_s = 150$ MeV) as described in Ref. [68]. A value of the bag constant $B^{1/4} = 180$ MeV is employed. (See Refs. [145, 68, 149] for a correct treatment of first order phase transitions in multi-component substances such as neutron star matter for which baryon and electric charge are the independent conserved charges.) Very little is known about the high-density properties of matter and our calculation does not imply a prediction of the rotational *frequency* or stellar *mass* at which a phase transition will occur. Rather it shows what the signal could be if conditions are attained for the phase change.

So far as we know, there is little difference in the properties of a neutron star that has no quark core and one that has already fully developed one. But a strong signal may be associated with the gradual conversion of matter from one phase into the other as the conversion is paced by the slow loss of angular momentum in the processes of electromagnetic radiation and electron-positron wind. The important observational features by which such an epoch could be identified are:

(1) The braking index has a value far from the canonical value, possibly by orders of magnitude and can be of either sign.

(2) The epoch over which the braking index is anomalous is long because

pulsars spin down slowly under ordinary circumstances but even more slowly when $|I'|$ is large (see (4.207)).

(3) The pulsar may be observed to be spinning up. (To avoid confusion with spin-up due to accretion, only isolated pulsars are relevant.)

(4) Except for the central part of the spin-up era, the derivative $\ddot{\Omega}$ is large and therefore easy to measure and so also is the braking index.

(5) Difficulty in measuring $\ddot{\Omega}$ can be used to deselect phase transition candidates.

(6) An estimated 1/100 pulsars may be passing through the transition epoch.

Pulsar observations are still in their infancy. It takes a considerable time-span of data to measure the braking index. And many of the presently known pulsars have only recently been discovered. We conclude from our work that it is plausible that the phase transition signal can be observed. It would be a momentous discovery to find that a phase of matter that existed in the early universe inhabits the cores of some neutron stars.

## 4.24   Neutron Star Twins

Could there exist in nature additional stable regions of dense stars beyond neutron stars? At first sight the answer seems to be negative. What else is there to stabilize a higher density sequence than the generic neutron star? If the quark constituents of baryons are deconfined as predicted by the theory of the strong interaction (QCD and its property of asymptotic freedom [150]), then baryon Fermi pressure is simply *replaced* (not supplemented) by the Fermi pressure of quarks. No apparent new agent for stabilization occurs as a consequence of deconfinement. Indeed, deconfinement will soften the equation of state by the loss of the short-range repulsive interaction between nucleons and by the sharing of nucleon Fermi pressure among three low-mass quark Fermi seas. Generally, deconfinement is expected to terminate the neutron star sequence when quarks are deconfined from a sufficiently large amount of hadronic matter.

Nevertheless, it is at least theoretically possible that a branch with higher density than normal neutron stars could exist. It may happen that, because of a phase transition, the equation of state is not sufficiently smooth as to obey the conditional theorem of Wheeler et al. [43]. Then there could exist non–identical twins of the same mass but different internal constitution than the canonical neutron stars [151]. We have investigated the possible existence of a third sequence of stable compact stars in the reference above and find that there are circumstances under which a third family does

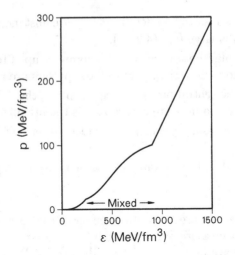

FIGURE 4.34. Equation of state of nuclear matter showing the pressure as a function of energy density in three phases: (1) low-density normal nuclear matter, (2) mixed phase of nuclear matter and quark matter, and (3) the high-density pure quark-matter phase. Neutron-star-matter has two independent components: they are baryon and electric charge [68]. The phase transition is first order. Consequently the pressure varies in the mixed phase rather than being constant, as in a one-component substance (such as pure water). Normal density of nuclear matter is 140 MeV/fm$^3$ and occurs in the first segment of this plot.

exist: They are non-identical *twins* of the normal family, being denser and therefore having a particle population corresponding to that higher density.

The equation of state for the sequence with maximum neutron star mass $\sim 1.42 M_\odot$, is shown in Fig. 4.34. The form is general for first order phase transitions of whatever origin in any substance with two or more independent components [68]. (For neutron stars the two independent components are electric and baryonic charge.) The three parts of the equation of state that are separated by discontinuities in slope correspond to (1) the pure quark-confined phase of nuclear matter, (2) the mixed coexistence phase of nuclear and quark matter, and (3) the pure deconfined quark matter phase.

Features of the stellar sequence shown in Fig. 4.35 can be identified with features in the equation of state shown in Fig. 4.34. One can see that near the end of the mixed phase $dp/d\epsilon$ becomes small; therefore also the adiabatic index, $\Gamma$, becomes small;

$$\Gamma \equiv d\ln p/d\ln \rho = (p+\epsilon)/p \cdot dp/d\epsilon. \qquad (4.210)$$

Pressure, energy density, and baryon density are denoted respectively by $p$, $\epsilon$, and $\rho$. In the upper region of the mixed phase, the pressure is too weak a function of energy density to sustain stability: consequently the canonical neutron star family terminates. The weakening of the adiabatic

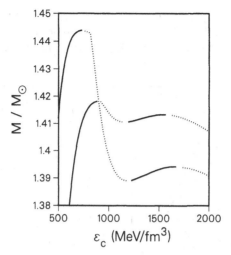

FIGURE 4.35. Two out of many examples of stellar sequences for which Neutron stars and a higher density stable family of 'non-identical twins' exist. In some cases, the denser family has a larger limiting mass than the neutron star family. Solid lines indicate the stable configurations. Stability is proven in connection with the discussion of Fig. 4.38.

index referred to is characteristic of first order phase transitions in substances having two independent components—in this case, baryon number and electric charge [68]. The adiabatic index is larger in the pure phases, and the pressure increase in the pure quark phase restores stability over a small range of central densities. We prove stability for our examples below. However, while the behavior of the adiabatic index described above is a quite general attribute for phase transitions in multicomponent substances, a stable third family is not. Stability is an integral property that depends, more or less, on the configuration of the bulk of the star and therefore on the equation of state over a broad density range. Consequently, the appearance of a third family will depend on the density region in which the critical behavior occurs.

## 4.24.1 PARTICLE POPULATIONS IN TWINS

We refer to stars of the denser sequence in Fig. 4.35 as non-identical twins of neutron stars because for both families it is the Fermi pressure of particles carrying baryon number that supports the star against gravity in addition to repulsion at short distance between any nucleons that may be present. In the present examples, both families contain deconfined quark matter, but in the first, only in the mixed phases. The particle populations are shown in Figs. 4.36 and 4.37 for a selected common mass. Both contain a region in which confined and deconfined matter coexist. The lower-mass twin has an 8 km radius core of *mixed* phase. whereas the inner 4 km of the higher-mass star is in the *pure* quark phase. As stressed above, however, it is not the particular content of the star that creates the additional family, but rather a particular way in which the adiabatic index changes with density.

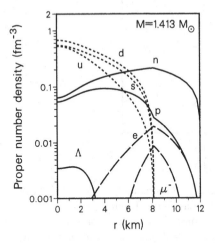

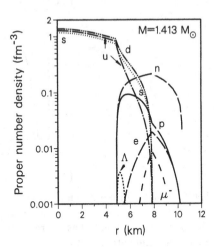

FIGURE 4.36. Particle populations of a lower-mass twin.

FIGURE 4.37. Particle populations in a higher-mass twin.

### 4.24.2   TEST FOR STABILITY

We now demonstrate stability to radial vibrations that would otherwise bring about collapse or explosion of a star. Stars on both segments of the stellar sequence shown in Fig. 4.35 that have positive slope $dM/d\rho_c > 0$ satisfy the necessary but *insufficient* condition for stability. Stability can be tested by an analysis of the radial modes of oscillation [91]. The squared frequency $\omega^2$ of the fundamental mode is plotted in Fig. 4.38. Positive values indicate stability and correspond to the segments with positive slope in Fig. 4.35. The analysis shows that the fundamental (nodeless $n = 0$) oscillation becomes unstable at the first maximum, as is usual, but unusually, stability of this mode is restored at the following minimum, to be lost again at the next maximum. The usual pattern is that the fundamental mode becomes unstable at the maximum in the neutron star family and a higher mode in order $n = 1, 2, 3 \cdots$ becomes unstable at each higher minimum and maximum. Undoubtedly the usual pattern resumes at densities higher than our third stable family. At such high densities that matter is in the pure quark phase, asymptotic freedom is likely to assure that the equation of state is smooth like a polytrope. Indeed, that is precisely asymptotic behavior of the MIT bag equation of state. From some density above the point where the equation of state is smooth, we are assured of the denumerable infinity of turning points and ever increasing number of unstable normal modes such as was found by Wheeler et. al. [43].

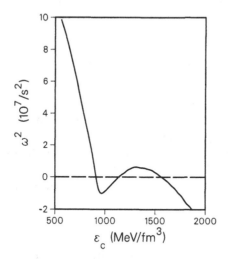

FIGURE 4.38. The square of the frequency of the fundamental ($n = 0$) radial vibration as a function of central density. Perturbations behave as $\exp(i\omega t)$ so they are unstable with diverging amplitude $\exp(|\omega|t)$ when $\omega^2 < 0$.

### 4.24.3  FORMATION AND DETECTION

From the above proof of stability we have shown that in principle a third family of stable degenerate stars could exist. Each or some of the stars on the high density branch have a non-identical twin on the low-density branch, having the same mass but different composition and radius. Can such twins be distinguished? One possible avenue is through observations on the so-called quasi-periodic oscillations in the X-ray brightness of accreting neutron stars. According to theory, mass and radius determinations may be possible [152, 153]. If twins exist, then the mass-radius curve will exhibit two segments of stable stars instead of one, and observed stars will fall on one or the other of the two distinct segments. The discovery of only about two stars on each branch with a radius resolution of a kilometer in our example would suffice to establish the existence of twin branches.

How could a high density twin be made in nature? The likely path to the high density twins is through the initial core collapse of a star, in which the core implodes through the normal star to the high density twin. Since no two supernova are likely to be identical, there being many variables that effect the outcome, like mass, rotation, symmetry, chemical composition of the progenitor, and the chaotic process of convection, it seems plausible that either twin could be produced. A second possible formation mechanism of the high density twin is through accretion onto a member of the low-density sequence of Fig. 4.35 followed by a minor explosion. Such explosions are known to occur in systems in which a white dwarf is in orbit with a normal star and accreting matter from it. These explosions are known as supernova type I.

## 4.25    Black Holes

"If there should really exist in nature any bodies, whose ... light could not arrive at us ... we could have no information from sight; yet, if any other luminous bodies should happen to revolve about them we might ... infer the existence of the central ones."
*Rev. John Michell in a letter to Henry Cavendish, May 26, 1783*
[109]

### 4.25.1    INTERIOR AND EXTERIOR REGIONS

So far as is known, the Rev. John Michell, an English philosopher and scientist and geologist of some repute in his own day, was the first person to conceive of a star from which light could not escape. A few years later, and presumably independently, the French scientist Pierre-Simon Laplace arrived at a similar notion. Michell reasoned that gravity should act on light as it does on mass. According to his understanding, stars might exist for which light would not have the escape velocity. We now know that the velocity of light is a constant; it is the energy (frequency) of light that is diminished in its flight from a star. Indeed, as we analyze the region of spacetime in and around black holes in this section, we will find, as shown in Fig. 2.1, that light cannot even travel in an outward direction if it is inside a black hole.

Nevertheless, Michell's conception was prescient. And as for the condition for confinement of light, consider the condition for the escape of a material body of mass $m$ and velocity $v$ from a star of mass $M$ and radius $R$,

$$\tfrac{1}{2}mv^2 > \frac{GmM}{R}\,. \tag{4.211}$$

Letting $v \to c$, we find that the critical radius is $R = 2GM/c^2$, which is the Schwarzschild radius of a black hole! (It is not clear to this authoe why classical reasoning should have led to an exact result of General Relativity.)

Now let us turn again to General Relativity. We have seen that Schwarzschild's solution to Einstein's equations in the region outside a static spherical star becomes singular at

$$r = r_S \equiv 2GM/c^2 \qquad (G = c = 1)\,. \tag{4.212}$$

If the star is enclosed by the Schwarzschild radius, it is a black hole. The spherically symmetric field described by Schwarzschild is that of a nonrotating black hole; it is often referred to as a Schwarzschild black hole.

The collapse of a star to a black hole was first discussed in the General Theory of Relativity by Oppenheimer and Snyder [38]. There are currently two classes of black holes of intense observational activity, objects of a few solar masses, the compact accreting partners of low-mass, X-ray binaries (LMXB) and extremely massive objects of $10^7$ to $10^{10} M_\odot$ in active galactic nuclei (AGN) (AGN stands for Active Galactic Nuclei, which are now understood to be galaxies, such as ours, at the center of which is a giant black hole that is ingesting stars from an accretion ring; as the stars are torn apart intense radiation is emitted.) The latter are probably ingesting stars of the surrounding galaxy by first reducing them to an accretion disk. In both cases, the radiation detected is thought to be produced by the accreting matter as it is heated by compression and friction while it spirals toward the hole. In the case of active galactic nuclei, radiation has been detected from X rays down to the infrared and probably includes gravitational radiation, though such has never been detected. The lighter black holes in binary systems are doing the same on a smaller scale. We discuss a third class of low-mass black holes and possible formation mechanisms in reference [4].

While not of primary concern to us here, it is nonetheless interesting to understand more fully the meaning of the singular point in the Schwarzschild metric. Therefore, we investigate the motion of a particle as it drops radially into a nonrotating black hole. Then two components of its four-velocity vanish, $u^2 = u^3 = 0$. The geodesic equation (3.38) for $u^0$—the equation of motion for a free particle in a gravitational field—is

$$\frac{du^0}{d\tau} = -\Gamma^0_{\alpha\beta} u^\alpha u^\beta = -2\Gamma^0_{10} u^1 u^0 = -2g^{00}\Gamma_{010} u^1 u^0 . \tag{4.213}$$

We used the symmetry of $\Gamma^\lambda_{\mu\nu}$ in its lower indexes (3.39) to arrive at the factor 2. Because $g_{\mu\nu}$ is symmetric, from (3.48) we have $\Gamma_{010} = \frac{1}{2} g_{00,1}$. Hence, we obtain

$$\frac{du^0}{d\tau} = -g^{00} g_{00,1} u^1 u^0 . \tag{4.214}$$

But

$$\frac{dg_{00}}{d\tau} = \frac{dg_{00}}{dr}\frac{dr}{d\tau} = g_{00,1} u^1 , \tag{4.215}$$

so recalling that for the special case of the Schwarzschild metric, $g^{00} = 1/g_{00}$, as found in (3.139), we arrive at

$$\frac{d}{d\tau}\left(g_{00} u^0\right) = g_{00}\frac{du^0}{d\tau} + u^0\frac{dg_{00}}{d\tau} = 0 , \tag{4.216}$$

Whence

$$g_{00} u^0 = C , \qquad (4.217)$$

where $C$ is a constant. Let the particle be dropped radially from rest with respect to the hole at $R > 2M$. Then $dr = d\theta = d\phi = 0$ at $R$ and, from the formula for the invariant interval (3.152), we have $u^0(R) \equiv dt/d\tau = 1/\sqrt{g_{00}(R)}$, so that

$$C = \sqrt{g_{00}(R)} = \sqrt{1 - \frac{2M}{R}} . \qquad (4.218)$$

From (3.4) we find

$$1 = g_{\mu\nu} u^\mu u^\nu = g_{00} (u^0)^2 + g_{11} (u^1)^2 . \qquad (4.219)$$

In the exterior region we have found $g_{00} g_{11} = -1$ from (3.152). Hence we have

$$g_{00} = C^2 - (u^1)^2 , \qquad (4.220)$$

or, using the explicit expression (3.150) for $g_{00}$,

$$u^1 = -\left( C^2 - 1 + \frac{2M}{r} \right)^{1/2} , \qquad (4.221)$$

where we have chosen the negative square root for the infalling particle. We now have expressions for $u^0$ and $u^1$ as functions of $r$. To get an expression between $t$ and $r$, we use

$$\frac{dt}{dr} = \frac{dt/d\tau}{dr/d\tau} = \frac{u^0}{u^1} = -C\left(1 - \frac{2M}{r}\right)^{-1}\left(C^2 - 1 + \frac{2M}{r}\right)^{-1/2} . \qquad (4.222)$$

We are interested in the behavior of $t$ close to $r_S$. Introduce a new variable $\rho$ defined by

$$r = 2M + \rho , \qquad (4.223)$$

with $\rho << 2M$, and study the behavior of $t$ for small $\rho$. Keep only leading terms in $\rho$:

$$\frac{dt}{dr} = -C\left(1 - \frac{2M}{2M + \rho}\right)^{-1}\left(C^2 - 1 + \frac{2M}{2M + \rho}\right)^{-1/2}$$

$$\approx -\frac{2M}{\rho} = -\frac{2M}{r - 2M} . \qquad (4.224)$$

Integrate to obtain

$$t = -2M \ln\left(\frac{r}{2M} - 1\right) + \text{const} . \qquad (4.225)$$

From this result we see that, as the particle approaches the Schwarzs-child radius $r \to r_S \equiv 2M$,      (upon setting $G = c = 1$) the time $t \to \infty$. An external observer at rest with respect to the star observes that the particle takes an infinite time to reach the Schwarzschild radius. For any process taking place on the particle (including the emission of light), the time to do so will appear to an exterior stationary observer to approach infinity as the particle approaches $r_S$. Light emitted from the particle will tend to be infinitely redshifted as seen by an exterior observer!

On the other hand, in a frame falling freely with the particle, we have $d\tau = dt$. So,

$$\frac{dt}{dr} = \frac{d\tau}{dr} = \frac{1}{u^1} = -\left(C^2 - 1 + \frac{2M}{r}\right)^{-1/2}. \tag{4.226}$$

As $r \to r_S$, $dt/dr \to -1/C$. An observer falling with the particle measures a finite time lapse for the particle to reach the Schwarzschild radius. Thus, there is no singularity in spacetime. Rather, the singularity is in the metric for which a transformation could be made to new coordinate system in which there is no singularity in the metric, and therefore no singularity in spacetime.

If the Schwarzschild radius lies outside a star, it separates space into distinctly different regions. From outside

$$r = r_S \equiv 2M, \tag{4.227}$$

(in units $G = c = 1$) one cannot see a particle drop into the hole. Neither can any particle or light escape from inside the Schwarzschild radius, as we shall see shortly. The Schwarzschild radius is sometimes referred to as the Schwarzschild horizon.

An external observer is unable to obtain any signal of what might happen within the Schwarzschild radius. Only the imprint of the mass is left on the external geometry in the case of a Schwarzschild black hole. In general two other attributes of a black hole leave their imprint on the external geometry: electric charge and angular momentum. The field produced by a rotating or charged star is different from that of Schwarzschild. We discuss rotating stars in [4].

### 4.25.2   NO STATICS WITHIN

To further analyze the nature of the nature of the Schwarzschild horizon, let us define a new time coordinate due to Eddington and Finkelstein whose form is suggested by (4.225):

$$t' = t + 2M \ln\left|\frac{r}{2M} - 1\right|. \tag{4.228}$$

Under this transformation, the Schwarzschild metric has no singularity as $r \to r_S$. [The second term cancels the singularity in (4.225)]. There are other coordinate transformations as well that are nonsingular at the horizon. We see that the Schwarzschild singularity is a coordinate singularity.

We now study the behavior of null cones with vertices at different spacetime points both far from, near, and inside $r_S$. The null cones (or light cones) are the hypercones in four-dimensional spacetime connecting an event at the vertex of a cone to neighboring events between which a light signal can propagate. Light rays travel into the future of the vertex event on those cones opening toward $+t$. Timelike geodesics ($d\tau^2 > 0$)—the paths followed by material particles that pass through the vertex of a cone—lie within the cone. Events in the past of the vertex event lie on or inside the cone opening in the opposite direction.

It will be sufficient to study the behavior of the cones in the t-r plane, taking $d\theta = d\phi = 0$. From the Schwarzschild expression for the proper time (3.152), the equation for the future cone of events at $r$ is

$$\left(1 - \frac{2M}{r}\right)dt^2 = \left(1 - \frac{2M}{r}\right)^{-1}dr^2 . \tag{4.229}$$

Transforming to the new time coordinate, we get

$$0 = \left(1 - \frac{2M}{r}\right)\left(\frac{dt'}{dr}\right)^2 - \frac{4M}{r}\left(1 - \frac{2M}{r}\right)\left|1 - \frac{2M}{r}\right|^{-1}\left(\frac{dt'}{dr}\right)$$
$$- \left(1 + \frac{2M}{r}\right). \tag{4.230}$$

The solutions are

$$\frac{dt'}{dr} = -1 \quad \text{and} \quad \left(1 + \frac{2M}{r}\right)\Big/\left(1 - \frac{2M}{r}\right). \tag{4.231}$$

So one side of the cone (closest to the star) always has slope $-1$ and the other has a slope that depends on the position of the event. We can write the equation for the latter as

$$(r_0 - r_S)\,dt' = (r_0 + r_S)\,dr , \tag{4.232}$$

where $r_0$ is used to define the $r$ coordinate at the vertex of the cone.

Far away from the Schwarzschild radius, the side of the cone that is remote from the black hole has slope $+1$. Therefore, the geodesic path of a light ray or particle that is far from the horizon can lead away or toward the black hole. These distant cones have axes parallel to the time axis or, more accurately, as the distance from the star approaches infinity they approach a parallel orientation. At event points closer to $r_S$, the cones tilt toward the star and have narrower opening angles. The cone with vertex at $r_0 = r_S$ has one side parallel to the $t'$ axis and the other pointing to smaller $r$, always with slope $-1$. Therefore a particle located interior to $r_S$, can only fall into the hole. Light emitted from the horizon can at best either circle the hole or fall in, but cannot escape.

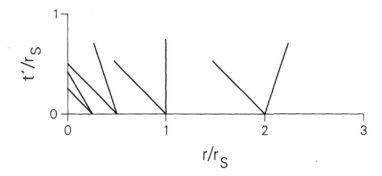

FIGURE 4.39. Future light cones at radial distances both inside and outside a black hole Schwarzschild radius $r_S \equiv 2GM/c^2$. Future points in spacetime of an event at the vertex of a cone can lie only within the cone; light can propagate on the cone itself. Inside the Schwarzschild radius, even light can propagate only inward.

The future cones of particles or photons already in the hole point only inward. There is no possibility of escape nor even of remaining at a stationary distance from the origin. Within the Schwarzschild radius, there is no possibility of statics or equilibrium! A star that falls through its Schwarzschild radius continues to collapse to a singularity or at least until the Planck scale[24] at which the classical theory may be invalid. (The Planck density is about $10^{78}$ times the central density of the most massive neutron star.)

To summarize, distant light cones have an opening angle approaching $\pi/2$ as the distance from the black hole increases. The cones tilt toward the hole for events that are closer and their opening angles become smaller. Inside the Schwarzschild radius, both sides of the cone point to a smaller radius: world lines can lead only inward. A few such light cones are illustrated in Fig. 4.39. It should be noted that they are not related to each other, but simply define the future light cone for each vertex point outside the black hole.

---

[24]Combinations of the Planck constant, the gravitational constant, and the speed of light define a Planck mass, length, and time. The first two define a density of $5 \times 10^{93}$ gm/cm$^3$. This can be compared to the central density of neutron stars $\sim 10^{15}$ gm/cm$^3$. Therefore, quantum gravity is of no relevance to compact, *stable* stars such as neutron stars and white dwarfs.

### 4.25.3   BLACK HOLE DENSITIES

Black holes are usually thought to have enormous densities. This is true of some. Those formed in the collapse of the degenerate cores of evolved stars have very high densities. Indeed, collapse continues to a singularity. We have seen that, within the horizon, all matter and even light can only fall inward. Collapse to the singularity is very rapid for solar mass stars, but slow for the collapse of a black hole having the mass of a galaxy. So it makes sense to enquire about the initial density of black holes of various masses.

Schwarzschild black holes are nonrotating stars whose radius satisfies $R < 2M$ $(G = c = 1)$. The average density satisfies

$$\rho = \frac{M}{(4\pi/3)R^3} \geq \frac{3}{32\pi M^2},\tag{4.233}$$

where (4.160) has been employed. Referring to the section on units (4.3), this can be expressed in the more convenient form,

$$\rho \geq \frac{1.8 \times 10^{16}}{(M/M_\odot)^2} \text{ gm/cm}^3.\tag{4.234}$$

The Milky Way galaxy has a mass of about

$$M_{\text{MilkyWay}} \approx 10^{12} M_\odot.\tag{4.235}$$

The density of a black hole formed from its collapse need only be larger than

$$\rho = 1.8 \times 10^{-8} \text{ gm/cm}^3.\tag{4.236}$$

The radius of the horizon is (recall that $M_\odot = 1.5$ km in gravitational units)

$$R = 3 \times 10^{12}\text{km},\tag{4.237}$$

which is about a third of a light year. This is very small compared to the dimension of the Milky way which is about 100,000 light years across.

### 4.25.4   BLACK HOLE EVAPORATION

Steven Hawking discovered [154] that, because of quantum effects, a black hole of mass $M$ radiates with a thermal spectrum of temperature,

$$T = (\hbar c^3)/(8\pi G k M),$$

where $k$ is the Boltzmann constant. The emission of this energy results in an energy decrease of the black hole, and thus a loss in its mass. It is interesting to calculate the interval of time $\tau$ that will be needed for a black hole to evaporate completely.

A black hole with mass $M$ has a Schwarzschild radius $R = 2GM/c^2$ so its surface area is $A = 4\pi R^2$ or

$$A = 16\pi G^2 M^2/c^4 .$$

Hawking radiation would have a power $P$ related to the hole's area $A$ and its temperature $T$ by the black-body law, $P = \sigma A T^4$ (Steffan–Boltzmann constant $\sigma$) which, gathering the above results yields

$$P = K/M^2 ,$$

$$K = (\sigma \hbar^4 c^8)/(256\pi^3 G^2 k^4) = 3.55 \times 10^{32} \text{ Watt} \cdot \text{kg}^2$$

But the power of the Hawking radiation is the rate of energy loss of the hole; therefore we can write $P = -dE/dt$. Since the total energy $E$ of the hole is related to its mass $M$ by $E = Mc^2$, we can rewrite the power as $P = -c^2 dM/dt$, so that

$$-c^2 dM/dt = K/M^2 .$$

Therefore,

$$M^2 dM = -[K/c^2]dt .$$

By integrating over $M$ from the initial mass $M$ to 0 and over $t$ from zero to $\tau$, we find the evaporation time,

$$\tau = [c^2/(3K)]M^3 \approx 2.11 \, (M/M_\odot)^3 \times 10^{67} \quad \text{yr} .$$

The mass of a black hole that will take the life of the universe of 15 billion years to evaporate is,

$$M = 8.9 \times 10^{-20} \, M_\odot$$

## 4.25.5  KERR METRIC FOR ROTATING BLACK HOLE

The field of a rotating star is not described by the Schwarzschild solution, but rather by the Kerr metric: The Kerr-Newman geometry is,

$$ds^2 = \frac{\Delta}{\rho^2}[dt - a\sin^2(\theta)\,d\phi]^2 - \frac{\sin^2(\theta)}{\rho^2}[(r^2 + a^2)d\phi - a\,dt]^2$$
$$- \frac{\rho^2}{\Delta}dr^2 - \rho^2\,d\theta^2 , \tag{4.238}$$

where

$$\Delta \equiv r^2 - 2Mr + a^2 + Q^2 ,$$
$$\rho^2 \equiv r^2 + a^2\cos^2\theta , \quad a \equiv J/M \tag{4.239}$$

and $a \equiv$ angular momentum per unit mass .

# 4.26   Problems for Chapter 4

1. Review how stars are made. Explain in your own words why stars of different mass evolve at different rates; why the final outcome is a white dwarf, neutron star or black hole; and why white dwarfs do not form a single one-parameter family of stars belonging to a unique equation of state. Why do neutron stars form a unique sequence? Write this information in a coherent interesting narrative.

2. Express the nucleon mass in grams, and divide into the Sun's mass to find the approximate number of nucleons contained in the Sun (recall Section 4.3). The Sun's binding energy is too small to be significant in this calculation.

3. On page 75 it was written that as the shock produced in the collapsing core of a star in its terminal stage moves outward, it disintegrates all nuclei in its path. Assuming as a round number that the binding energy per nucleon in a nucleus is 10 MeV, and having learned that there are several times $10^{57}$ nucleons in a solar mass and that the core mass is approximately the Chandrasekhar mass, show that the energy dissipation is several times $10^{51}$ ergs. This recall (see page 75) is the same order of magnitude as that of the kinetic energy of the material that will finally be ejected in the supernova. Thus the failure of the *prompt* bounce scenario of supernova.

4. Given the Sun's mass in grams, as in (4.10) calculate its mass express-ed in gravitational units, that is, in kilometers. What is the Schwarzs-child radius of the Sun?

5. On page 81 the moment of inertia of the Crab pulsar was quoted as 70 km$^3$. Given a uniform density object of a solar mass and 10 km radius, what is its classical moment of inertia in the above units and in g cm$^2$.

6. Write computer programs to calculate the equation of state for an ideal gas of neutrons, protons and electrons in a charge neutral ad-mixture (4.85) in the ground state at whatever density (4.83). As re-marked at the beginning of Section 4.9.3, the exact equations (4.79) cannot give accurate answers on a computer of finite word length when the momentum is small compared to particle mass (different for nucleon and electron) and vice versa. You will need the formu-lae (4.79) to compute energy density, pressure and particle number density for all three particle types. But the answers computed will be inaccurate, at different densities, for nucleons and electrons. There-fore, the *exact* algebraic expressions will have to be supplemented by the approximate expressions derived in Section 4.9.3 at high or low density. Be aware that the limits refer to $k/m$ or its inverse so that the

approximate expansions will provide more accurate computer results than the exact algebraic expressions at different *baryon* density for nucleons than electrons. Compute the equation of state from below neutron threshold density, (4.88), say, $\rho \equiv \rho_n + \rho_p = 1 \times 10^{-9}$ fm$^{-3}$ to 10 times the density of normal nuclear matter ($\rho_0 \sim 0.15$ fm$^{-3}$). Compare the *exact* algebraic and the approximate numerical results. You will probably notice that the exact algebraic formula are not even computable over certain ranges and will have to place conditions in your program to avoid overflows or underflows. Compare your final answer with Fig. 4.8.

7. Verify the expression for the pressure (4.76) by first deriving the explicit expressions (4.80) and then performing the volume derivative of the energy. This type of expression will occur frequently.

8. Make sure to understand the derivation of the equilibrium condition (4.83). Such expressions recur frequently.

9. Derive the Fermi gas expression for $k_p$ in terms of $\rho$, (4.92). Confirm the limiting proton fraction in this model.

10. Derive (4.95) and (4.95).

11. Derive the low–density approximations (4.96).

12. Verify equation (4.101). Follow the logic that leads to the conclusion that the mass of a white dwarf resides principally in baryons, but it is electron pressure that supports it against collapse.

13. Verify (4.107) from the results preceding the equation and hence confirm the validity of the Newtonian approximation for white dwarfs.

14. Derive the Newtonian equation (4.108).

15. Carry out the Lane–Emden transformation of the Newtonian equation. Verify the expressions for mass and radius and their relationship. For $\gamma$ precisely 4/3 what is the radius?

16. Verify that the polytrope equation of state written in Section 4.9.6 for the ultrarelativistic electron region. Write a computer program to solve the Lane–Emden equation. Verify that the solution for $\gamma = 4/3$ has a zero at the value of the argument stated in the above section. Confirm the Chandrasekhar mass (4.125).

17. How could one infer the contents of a white dwarf, such as carbon or magnesium or any other element?

18. Verify the values of mass and radius of a purely neutron star that were first obtained by Oppenheimer and Volkoff, as quoted in Section 4.9.8 by writing a computer program to solve the Oppenheimer–Volkoff equations. You will first have to compute the equation of state of pure neutron matter. Derive a table of $\epsilon, p$, and $\rho$. The equations of stellar structure are of first order and can be solved by a simple difference procedure, starting at the origin, computing $dp$ and $dM$ at the next interval $0 + dr$, and so on. Repeat for a number of values of the central energy density and thus develop a stellar sequence of mass as a function of central density. Express the mass in units of the solar mass. If you have transformed energy density into $1/\text{km}^2$ as in (4.9), then an integration over distance in units of km will produce a mass in units of km. The solar mass is given by (4.10). Test the accuracy of the answers by repeating the calculation for several choices of step size $dr$.

19. Compute the lattice energy and pressure (4.138). Hence derive the equation of state (4.141) for white dwarf matter consisting of a single nuclear species $A, Z$ plus electrons.

20. Assume the distance $dr$ from a neutron star surface at which the energy density has risen from its surface value to $1 \cdot 10^6$ gm/cm$^3$ is small compared to the radius $R$ of the star. For $R = 10$ km and $M = 1.5 M_\odot$ estimate $dr$ using (3.180) and Table 5.6 of [4]. Find that $dr \approx 76$ cm.

21. Why might the Schwarzchild radius be obtained from the classical condition (4.211)?

22. Give details of the proof of (4.218) starting at the geodesic equation for $u^0$.

23. Prove (4.222), (4.224), and (4.225) to see that an external observer would measure an infinite time for a freely falling particle to approach the Schwarzschild horizon.

24. Prove that in a freely falling frame, $dt/dr \longrightarrow 1/C$ as $r \longrightarrow r_S$—that an object will fall into a black hole with a finite elapse of time.

25. Derive the equations defining the slopes of the two sides of the future light cone (4.231).

26. Find the lower limit of black hole density (4.234). Confirm the numbers quoted below the formula for the critical density of the Milky Way and the Schwarzschild radius. Estimate the actual average density of the galaxy.

# 5

# Cosmology

## 5.1 Foreword

Cosmology is the science of the evolution of the universe. Einstein's theory of General Relativity provides the framework in which the likely, and in many respects verifiable, theory of that evolution can be traced. There are a number of good books on the subject. The purpose here is to give a brief overview of cosmology. Einstein, himself, used his theory in this connection. But the time was before Edwin Hubble's amazing discoveries about our universe—its expansion with distant galaxies receding in proportion to their distance from us, and very importantly, that on the large–scale, the universe is homogeneous and isotropic. These observations make it clear that we are not at the center of the universe; rather, the universe is expanding *from everywhere.*

There is overwhelming evidence that the universe—after the beginning—was immensely hot and dense. The very fact that the universe is expanding now, and is pervaded by a low-temperature radiation that is uniform on the sky—not associated with any particular galaxy—implies that it was once smaller, denser, and hotter. Then too, the primordial elements—hydrogen and helium with a little deuterium—could have been synthesized only at the enormous particle density and temperature that prevailed then. The expansion cooled the universe too rapidly to allow the formation of elements heavier than helium and traces of other light ones. The remainder had to await the formation of the first stars in which successive burning stages combined hydrogen into helium, three and four helium to form carbon and oxygen, and so on. It was only in the demise of stars in supernova explosions that their store of elements are released into the cosmos, while the heavy elements (greater than iron) were produced in the maelstrom outside the exploding star.

In recent years there has been tremendous progress in experimental cosmology. The very seeds of the future galaxies of stars are found in the cosmic background radiation (CMBR). And the age of the universe has been determined to be 13.7 billion years within about a percent. The cosmological constant that Einstein introduced into his theory because such a constant was allowed in his field equations, was adjusted by him to prevent the universe from expanding *or* contracting under gravity. In those days it was a fixed belief in Western thought that the universe existed as it was from time everlasting. Only after his introduction to Hubble in California, did Einstein realize that with his theory he could have predicted the expansion that Hubble discovered—even an accelerating expansion—as has been discovered in our own day by independent teams led by Saul Perlmutter and by A. Reiss. Much earlier, in the 1920's a Russian bomber pilot and meteorologist, Alexandre Friedmann, and a Belgian Priest turned cosmologist, Georges Lemaître, had studied Einstein's theory and independently concluded that the universe was not static but would possibly recollapse, or expand indefinitely, maybe even with acceleration.

## 5.2   Units and Data

$c \approx 3 \times 10^5$ km/s   (speed of light)
$G$ (Newton's constant) $\approx 6.7 \times 10^{-8}$ cm$^3$/(g s$^2$)
$y \approx 3.2 \times 10^7$ s   (year in seconds)
Distance : ly (lightyear) $= 9.5 \times 10^{12}$ km
Distance : pc (parsec) $= 3.3$ ly $= 3.1 \times 10^{13}$ km
$H_0$ (Hubble constant) $\approx 200$ km/s $\cdot 1/(10^7$ ly)
$1/H_0 \approx 15 \times 10^9$ y $\approx 4.6 \times 10^{17}$ s
parsec $\approx 3.3$ ly $\approx 3.1 \times 10^{13}$km
Universe age $\approx 1/H_0$
Earth age $\approx 4.5 \times 10^9$ y
Earth mass $\approx 6 \times 10^{27}$ g
Sun age $\approx 4.5 \times 10^9$ y
Sun mass $M_\odot \approx 2.0 \times 10^{33}$ g
Milky Way age $\approx 10^{10}$ y
Milky Way mass $\leq 10^{12} M_\odot$
Life of a 10M$_\odot$ star $\approx 10^7$ y
Time since dinosaurs $\approx 7 \times 10^7$ y
' $\approx$ ' means accurate to two figures

## 5.3   World Lines and Weyl's Hypothesis

Time always progresses, and in one direction only. Therefore, even if a particle sits still, nevertheless, in four-dimensional spacetime it traces a track called its world line. A mathematician, Herman Weyl, hypothesized that world-lines of particles in a universe such as ours do not become entangled.[1] In a uniform and expanding universe, an observer could see the world-lines diverging from a  point at some distant finite or infinite time

---

[1] If any two world lines were to cross, the single-valuedness of functions of time would be lost.

in the past, but never again would they meet.[2] This is an altogether reasonable idealization. Of course, sometimes galaxies collide. But as to the overall history of the universe, we are not disturbed by these events.

If on each world-line a common *time t* is marked, the points so singled out form a *spatial* surface. The surface might simply be a plain in which case, geometry on this plain would be Euclidean, like the geometry of lines inscribed on a sheet of paper on a desk. In fact, recent discoveries in cosmology confirm that this is indeed the curvature of our universe—flat. It need not have been so. The surfaces at any slice of cosmic time could have been spherical or hyperbolic.

From what has been said, it is not obvious that there exist only three possibilities for the curvature of the universe. The three possibilities mentioned emerge as the only curvatures that are described by the metric of a uniform homogenous universe. Robertson and Walker discovered this metric independently. Cosmology adopts Weyl's hypothesis concerning world lines; they do not become entangled. In such a case, we may choose a point on each world line of all the particles or galaxies in the cosmos that carry the same label $t$. These points define a surface in 4-space. They are parameterized by $t$, which is called cosmic time. When we speak of time, or cosmic time in cosmology, one is referring to events on such a surface.

## 5.4    Metric for a uniform isotropic universe

We have previously introduced the Robertson–Walker metric. Here we derive it. Express Weyl's hypothesis in terms of coordinates and metric. A world line is labeled by three space coordinates $x^m$ ($m = 1, 2, 3$) and a time coordinate $x^0$. Consider a 3-surface defined by an orthogonal slice through the world lines at a common time $x^0$ which we use to label such slices. To satisfy the Weyl hypothesis, the metric tensor $g^{mn}$ must have the following properties. Orthogonality is expressed by $g_{0n} = 0$. Each of the world lines $x^m = $ constant, is a geodesic. Therefore,

$$\frac{d^2 x^m}{ds^2} + \Gamma^m_{kl} \frac{dx^k}{ds} \frac{dx^l}{ds}$$

where the line element is $ds^2 = g_{kl} dx^k dx^l$. For $x^m = $ constant (each $m = 1, 2, 3$) we obtain $\Gamma^n_{00} = 0$, and $\partial g_{00}/\partial x^n = 0$. Thus $g_{00}$ depends only on $x^0$; we can therefore replace it by a suitable function of itself that makes $g_{00} = 1$. The line element then becomes $ds^2 = c^2 dt^2 + g_{mn} dx^m dx^n$ where $t \equiv x^0$ is cosmic time.

---

[2]It is remarkable that Weyl introduced his hypothesis before Hubble had discovered the expansion of the universe.

1. Example: Surface of negative curvature

$$x_i^2 - (ct)^2 = -R^2 \qquad (5.1)$$

Substitute

$$x_1 = R\sinh\chi \, \cos\theta, \quad x_2 = R\sinh\chi \, \sin\theta \, \cos\phi \qquad (5.2)$$

$$x_3 = R\sinh\chi \, \sin\theta \, \sin\phi, \quad t \equiv x_4 = R\cosh\chi \qquad (5.3)$$

This gives,

$$dx_i^2 - (cdt)^2 = R^2 \left(d\chi^2 + \sinh^2\chi(d\theta^2 + \sin^2\theta d\phi^2)\right) \qquad (5.4)$$

Now substitute $r = \sinh\chi$ to obtain

$$c^2 dt^2 - R(t)^2 \left( \frac{dr^2}{1 + kr^2} + r^2(d\theta^2 + \sin^2\theta \, d\phi^2) \right) \qquad (5.5)$$

This is known as the Robertson–Walker metric. The constant $k$ is called the curvature parameter; it can take three values, $k = 1$ for spherical subspace, $k = 0$ for a planar, and $k = -1$ for a hyperbolic. Note that the spatial part is symmetric with a universal curvature characterized by $R(t)$. That is the sole degree of freedom in a uniform homogeneous universe.

## 5.5    Friedmann—Lemaître Equations

It is not really necessary to understand General Relativity, to study cosmology, if one takes on faith that the universe is uniform and isotropic—true on the large scale—as is evident from the large sky surveys of galaxies. In that case the general metric takes a particular from named after Robertson and Walker (see section 5.4). It is,

$$ds^2 = c^2 dt^2 - R(t)^2 \left( \frac{dr^2}{1 + kr^2} + r^2(d\theta^2 + \sin^2\theta \, d\phi^2) \right) . \qquad (5.6)$$

The constant $k$ is called the curvature parameter; it can take three values, $k = 1, 0, -1$ for a spherical, planar, and hyperbolic subspace. Note that the spatial part is symmetric with a universal scale characterized by $R(t)$. The scale factor, $R(t)$, is the sole degree of freedom in a uniform homogeneous universe, and represents the size in unspecified units. It is of prime importance.

For the Robertson-Walker line element corresponding to a *homogeneous and isotropic* universe, only two of Einstein's field equations are independent. They can be taken as

$$\dot{R}^2 + kc^2 = (1/3)(\Lambda + 8\pi G\rho)R^2 , \qquad (5.7)$$

and

$$\ddot{R} = (1/3)[\Lambda - 4\pi G(\rho + 3p/c^2)]R. \tag{5.8}$$

These are called the Friedmann and Lemaître equations after the men who derived them from Einstein's theory. Here, $\rho$ is the mass density of matter and radiation, $\epsilon = \rho c^2$ the corresponding energy density, $p$ the pressure, $\Lambda$ is Einstein's cosmological constant, and $k$ is the curvature parameter of the universe.

Several interesting results can be derived. Take the derivative of the first of the above pair, multiply the second by $\dot{R}$ and eliminate the $\Lambda$ term from the resulting pair to obtain the conservation law implicit in the Einstein equations (divergenceless stress-energy tensor),

$$\dot{\rho} = -3(p/c^2 + \rho)(\dot{R}/R). \tag{5.9}$$

This equation can also be written in two different ways:

$$d/dt(\rho c^2 R^3) = -p\, dR^3/dt, \tag{5.10}$$

which is the energy-work equation for expansion or contraction. Another way in which the conservation equation can be written is

$$d\rho/dR = -3(p/c^2 + \rho)/R. \tag{5.11}$$

The independent equations governing expansion may be taken as the first of the Friedmann–Lemaître equations together with the local conservation equation in any of its forms.

We can derive rigorously the behavior of radiation and matter densities as the universe expands from either of the conservation equations. The equation of state for radiation is $p = (1/3)\rho_r c^2$. Therefore

$$d\rho/\rho = -4dR/R. \tag{5.12}$$

This yields the conservation equation

$$\rho_r \sim 1/R^4, \tag{5.13}$$

for radiation. For matter, $p \ll \rho_m/c^2$ and we obtain instead

$$\rho_m \sim 1/R^3. \tag{5.14}$$

Thses equations are important: they tell us that radiation dominated early in the history of the universe, and matter next. Finally after radiation and matter had become diluted because of the expansion, the cosmological constant became the dominating factor in the subsequent course of the universe as can be seen from either of the Friedmann–Lemaître equations 5.7 or 5.8.

## 5.6    Temperature Variation with Expansion

We derived above the variation of the densities of radiation and matter with universal expansion. Here we give a qualitative account of the temperature variation. The wavelength of radiation $\lambda$ is stretched by the expansion according to

$$\lambda_0/\lambda = R_0/R, \qquad (5.15)$$

just as the wavelengths of photons in a box are stretched if the box dimensions are increased. The number of photons is conserved because their number is $2 \times 10^9$ (section 5.16) times the number of baryons so that photon scattering is extremely rare. Scattering becomes impossible, except for photon-photon scattering, after electrons have combined with protons and nuclei, referred to as decoupling. (This event is sometimes inappropriately referred to as recombination.) It follows that $N \sim 1/R^3$ so that the energy density

$$\epsilon(\nu) = \sum h\nu N(\nu), \qquad (5.16)$$

transforms as

$$\epsilon/\epsilon_0 = (R_0/R)^4. \qquad (5.17)$$

The spectral distribution depends on $h\nu/T$ so that $T$ is altered in the same way as $\nu$, namely

$$T/T_0 = R_0/R. \qquad (5.18)$$

## 5.7    Expansion in the Three Ages

The expansion of the universe is controlled by Einstein's equations. For a uniform homogeneous universe we have seen that they take a simple form, reducing to two in number, the Friedmann–Lemaître expansion equation and the continuity equation for conservation of energy (page 190). During the radiation age, earlier than about a million years, the expansion equation takes the simple form;

$$\dot{R}^2 = 8\pi G\rho_0 R_0^4/3R^2, \qquad (5.19)$$

where $\rho_0$ and $R_0$ are the values of the density and scale factor at any convenient reference time (for example the present). From this equation we learn that the size of the visible universe increases with time in proportion to the square root of time,

$$R \sim \sqrt{t}, \qquad (5.20)$$

but ever more slowly ($\dot{R} = 1/2\sqrt{t}$). Meanwhile, the temperature falls as

$$T \sim 1/R \sim 1/\sqrt{t}. \tag{5.21}$$

The mass densities of radiation and matter became equal at about a million years; radiation slowly faded thereafter. Equality marked the beginning of the *second age* when the universe was dominated by matter. When $\rho_m$ is the dominant term of those on the right of the Friedmann–Lemaître equation, the universal expansion is controlled by

$$\dot{R}^2 = 8\pi G\rho_0 R_0^3/3R. \tag{5.22}$$

In this age, the expansion increases with time as $R \sim t^{2/3}$; the speed of expansion continues to decelerate because of the gravitational attraction of matter and radiation. This is the age we have recently (several billion years ago) passed out of.

However, there is a *third age* when density has diluted and the dark energy term $\Lambda$ dominates. Let us see what its effect is; the Friedmann–Lemaître equation becomes with time

$$\dot{R}^2 = \Lambda R^2/3. \tag{5.23}$$

The solution to the expansion equation for positive $\Lambda$, the dark energy, is

$$R \sim \exp\sqrt{\Lambda/3}\,t. \tag{5.24}$$

We can also note that

$$\ddot{R} = (\Lambda/3)R, \tag{5.25}$$

so that the universal expansion *accelerates* in the third era.

We can summarize the time dependence of expansion as follows;

$$1 + z = \frac{R_0}{R} = \begin{cases} (t_0/t)^{1/2} & \text{radiation era} \\ (t_0/t)^{2/3} & \text{matter era} \\ \exp\left[\sqrt{\Lambda/3}\,(t_0 - t)\right] & \text{dark energy era} \end{cases}$$

In the first two ages, the expansion decelerates. In the third it accelerates. The acceleration in the three ages is summarized as;

$$a \sim \begin{cases} -1/(t)^{3/2} & \text{radiation era} \\ -1/(t)^{4/3} & \text{matter era} \\ \Lambda/3\exp\left[\sqrt{\Lambda/3}\,t\right] & \text{dark energy era} \end{cases}$$

We may summarize as follows: In the first two eras, gravity, acting on radiation and matter decelerated the expansion, but differently, according to which dominated the contents of the universe. In the third era —the era we are in now—dark energy represented by the cosmological constant, has become dominant, giving rise to an accelerating expansion, presumably forever.

## 5.8   Redshift

In the context of an *expanding* universe what was originally called the Doppler shift is more accurately named the *cosmological redshift*. The redshift of radiation from distant galaxies is caused by the elongation of light wavelengths in an *expanding* universe in which the galaxies are co-movers. The Doppler shift refers to radiation received by an observer from an object moving with respect to him, and both in an unchanging space. True, the universe is expanding, but galaxies, the solar system and so on, are not expanding with the universal expansion. They are gravitationally bound entities that are co-movers in the expanding universe.

In cosmology, a past event is usually referenced by the value of the redshift $z$, the fractional change in wavelength of emitted and received light, because that is what can be measured, whereas the time at which it occurred cannot. An approximate time can be referenced only with an assumption of how the universe actually evolved with time, which of course is not measurable. The best one can do is to use a model of the expansion by reference to a particular scenario, say the Friedmann and Lemaître equation with definite assumptions about the cosmological parameters that appear in it. These parameters $k$, $\Lambda$, and the densities of matter and radiation $\rho$ are known only within errors and there are three of them against one, the redshift $z$.

The redshift of light emitted by a receding source, say a galaxy, is the fractional change in wavelength of light between that received by an observer $\lambda_o$ and that emitted by the source $\lambda$,

$$z = (\lambda_o - \lambda)/\lambda.$$

Because of the scaling of wavelength with expansion it follows that

$$z = (R_o - R)/R \equiv \Delta R/R = (\Delta R/(\Delta T R))\Delta T \approx (\dot{R}/R) \cdot \Delta T$$

$$\equiv H\Delta T.$$

Where $H$ is the Hubble constant,

$$H = \dot{R}/R,$$

and $\dot{R} = dR/dt$. The distance to source is $R = c\Delta T$, while the Hubble law gives $v = HR$. So $z = (v/R)(R/c)$ or $z = v/c$ approximately for not too distant galaxies. Why not too distant? Because the expansion velocity was not a constant over the long term. At first the expansion was decelerated by gravity, and later accelerated by dark energy.

## 5.9   Hubble constant and Universe age

The measurement of the Hubble constant is difficult, and various means give somewhat different answers. It is an ongoing effort to accurately determine it. Hubble estimated it by plotting the distance of nearby galaxies as determined by parallax (good to 100 ly ($= 3 \times 10^{15}$ km) versus the velocity of recession as determined by the redshift or Doppler shift. Extrapolation to larger distance is not reliable and other methods have to be used. A current estimate of the Hubble constant is

$$H_0 = 200 \ \frac{\text{km}}{\text{s}} \ \text{per} \ 10^7 \ \text{ly} \,.$$

This yields for the universe age estimated as the inverse of Hubble, $1/H_0 = 15 \times 10^9$ years.

The Hubble constant is poorly known because a measurement can be made only at the present whereas, the expansion need not have been at uniform speed and almost certainly was not. After all, gravity slowed the expansion for much of the life of the universe, and dark energy, represented by Einstein's cosmological constant implies a current acceleration which began possibly 2 By ago. The age of the universe as given by the Hubble constant is therefore only approximate. The age can only be inferred from astronomical observations of old stars, globular clusters and the oldest (therefore coldest)white dwarfs. The latter method presently yields an estimate of about 12.6 billion years, while some observers claim 13.7 By.

## 5.10   Evolution of the Early Universe

Of the three terms on the right side of the Friedmann–Lemaître equation (page 190), only the one containing the density of mass, $\rho$, is important at early times. Therefore the universal expansion is governed by

$$\dot{R}^2 = (8\pi G\rho(t)/3)\, R^2(t) \,. \tag{5.26}$$

Radiation dominates in the early universe so from the Stefan–Boltzmann the equivalent mass density is:

$$\rho(t) = aT^4(t)/c^2 \,, \tag{5.27}$$

where $k$ is the Boltzmann constant, and

$$\begin{aligned} a &= 8\pi^5 k^4/15(ch)^3 \\ &= 7.57 \times 10^{-15} \ \text{g}/(\text{cm s}^2 \ ^\circ\text{K}^4) \,. \end{aligned} \tag{5.28}$$

The wavelength of radiation is Doppler shifted by the expansion so that $T(t) \sim 1/R(t)$ and consequently

$$\dot{R} \sim -\dot{T}/T^2 \,. \tag{5.29}$$

Therefore the Friedmann–Lemaître expansion equation can be written as a time evolution equation for the temperature:

$$\dot{T} = -\sqrt{8\pi Ga/3c^2}\, T^3(t) \,. \tag{5.30}$$

The solution is

$$T^2 = \sqrt{3c^2/32\pi Ga} \, / \, t \, . \tag{5.31}$$

Therefore, in the early universe, temperature decreases as $T \sim 1/\sqrt{t}$, radiation density decreases as $\rho \sim 1/t^2$, and the universe expands as $R \sim \sqrt{t}$. In particular

$$R(t) = T_0 R_0 (32\pi Ga/3c^2)^{1/4} \sqrt{t} \tag{5.32}$$

where $T_0$ is the present temperature of the background radiation and $R_0$ is the present scale factor, which can be taken as unity. Inserting the fundamental constants we find for the factor appearing in the above equation:

$$\sqrt{3c^2/32\pi Ga} = 2.31 \times 10^{20} \text{ s } {}^\circ\text{K}^2 \, . \tag{5.33}$$

We emphasize that the above solution for $T$ holds only in the early universe when radiation was the dominant component of energy. We can also note that the radiation density scales as

$$\rho \sim T^4 \sim 1/R^4 \, . \tag{5.34}$$

## 5.11  Temperature and Density of the Early Universe

The results of Section 5.10 took account only of photons. It is trivial to improve those results. At high temperature when $kT$ is large compared to particle masses, the equilibrium number of particles $(N)$ and antiparticles $(\bar{N})$ in a vacuum or the early universe is given by statistical mechanics as

$$N = \bar{N} = 4\pi g/h^3 \int_0^\infty p^2 \, dp/(e^{pc/(kT)} \pm 1) \tag{5.35}$$

where $g$ is the statistical weight of the species, the $(+)$ sign holds for Fermions and the $(-)$ sign holds for Bosons. The following results apply. Photons and Bosons:

$$N = 0.488 \, x^3 \text{ meter}^{-3}, \qquad \epsilon = aT^4 \, . \tag{5.36}$$

Electrons (each flavor), Nucleons, Hyperons and their antiparticles:

$$N = \bar{N} = 0.183 \, gx^3 \text{ meter}^{-3}, \qquad \epsilon = (7/8)g \, aT^4 \, . \tag{5.37}$$

Neutrinos and Antineutrinos (each flavor):

$$N = \bar{N} = 0.091 \, x^3 \text{ meter}^{-3}, \qquad \epsilon = (7/16)g \, aT^4 \, , \tag{5.38}$$

where $x = 2\pi kT/(hc)$, and $a$ is the Stefan–Boltzmann constant. The total energy density therefore has the form

$$\epsilon = \alpha(T)aT^4 . \tag{5.39}$$

To take account of all these species, instead of merely photons, the results of Section 5.10 should be modified by the substitution $a \rightarrow a\alpha(T)$. For example, at an epoch during which the temperature is greater than the electron mass but less than the muon mass, that is when $10^{12} > T > 5 \times 10^9$ and there are photons, electrons, positrons, and 3 flavors of neutrinos and their antineutrinos, the degeneracy is

$$\alpha = 1 + 2 \times 7/8 + 2N_f \times 7/16 = 43/8 . \tag{5.40}$$

However, when $T < m_e = 5 \times 10^9$ degrees, the degeneracy factor became $\alpha = 29/8$ (photons, and 3 flavors of neutrino pairs). Using the results of Section 5.10 we obtain the convenient connection between time and temperature in the radiation-dominated universe;

$$T = 1.5 \times 10^{10}/(\alpha^{1/4}\sqrt{t}) , \tag{5.41}$$

where $T$ is in K$^\circ$and $t$ is in seconds. Another useful relation gives the mass density of radiation,

$$\rho = aT^4/c^2 = 4.5 \times 10^5/(\alpha t^2) \text{ g/cm}^3 . \tag{5.42}$$

## 5.12   Derivation of the Planck Scale

The condition that the de Broglie wave length of the visible universe fits within the cosmic horizon defines the time we seek,

$$h/p = \lambda = ct . \tag{5.43}$$

In a volume of such dimension, the number density of particles and photons behaves as

$$n \sim 1/\lambda^3 . \tag{5.44}$$

In this confined region, particles are ultra relativistic so that their energy is

$$E \sim pc = hc/\lambda . \tag{5.45}$$

Consequently the equivalent mass density is

$$\rho \sim nE/c^2 = h/(\lambda^4 c) = h/(c^5 t^4) . \tag{5.46}$$

At high density, the Friedmann–Lemaître equation simplifies and yields

$$\rho \sim 1/(Gt^2). \tag{5.47}$$

Combining the last two equations we find that

$$t \sim \sqrt{hG/c^5} \approx 5.4 \times 10^{14} \text{ sec}, \tag{5.48}$$

which is the time we sought, the Planck time, when the visible universe lay within its de Broglie wavelength.

The other Planck quantities, distance, mass, and density are various combinations of the same fundamental constants.

## 5.13    Time-scale of Neutrino Interactions

Neutrinos decouple from the rest of the universe when the mean time between interactions with matter exceeds the age of the universe. To calculate the age of the universe at that time, we first compute the interval between interactions.

The density of neutrinos varies as

$$N \sim 1/R^3 \sim T^3 \tag{5.49}$$

because $T \sim 1/R$ where $R$ is the scale factor of the universe. The cross-section for the weak neutrino reactions is

$$\sigma \sim G_A^2 (kT)^2, \tag{5.50}$$

so the time scale for the reactions is

$$\tau \sim 1/(\sigma N c) \sim 1/T^5. \tag{5.51}$$

where $G_A$ is the weak-interaction coupling constant. By contrast, the scale factor varies as $R \sim 1/T$, so that the time between reactions very soon becomes longer than the age of the universe. Neutrinos drop out of equilibrium at one second.

## 5.14    Neutrino Reaction Time-scale Becomes Longer than the Age of the Universe

By equating the interval between neutrino interactions with matter computed in Section 5.13 with temperature a function of time as derived in Section 5.10, we obtain thereby an equation whose solution is the instant when the two times are equal. If we had actually computed the constant of proportionality, we would find that instant to be $t = 1$ second. We can calculate the time after which the interval between neutrino reactions (see Section 5.13) became longer than the age of the universe. Both the neutrino reaction time-scale and life of the universe are functions of temperature (Section 5.10). By eliminating temperature we find immediately

$$t \sim G_A^{4/3}(3c^2/32\pi G\chi a)^{5/6} . \tag{5.52}$$

After this time the reaction rate is smaller than the expansion rate of the universe so there are no neutrino reactions at all.

## 5.15  Ionization of Hydrogen

The ionization energy of hydrogen is 13.6 eV. From the conversion

$$kT = 8.63 \times 10^{-5} \text{ eV} \tag{5.53}$$

this energy corresponds to a temperature of $1.5 \times 10^5$ Kelvin. Yet, at much lower temperature, hydrogen was ionized. The point is that the tail of the Planck distribution, *because* of the very high ratio of photons to baryons $(2 \times 10^9)$ from Section 5.16, contained a large number of ionizing photons.

## 5.16  Present Photon and Baryon Densities

From the present temperature of the radiation background, $T_0 = 2.728$ K° , the *present* mass density of radiation can be calculated from Boltzmann's law,

$$\rho_r(t_0) = aT_0^4/c^2 = 4.66 \times 10^{-34} \text{ g/cm}^3 . \tag{5.54}$$

The present *number* density of photons (i.e. the number per cubic centimeter) can be found from Planck's law for blackbody radiation and the measured cosmic background temperature. The number density is

$$n_\gamma(t_0) = 0.244(2\pi kT_0/hc)^3 = 413 \text{ photons/cm}^3 . \tag{5.55}$$

From the baryon to photon ratio it is known that $n_B/n_\gamma = 5 \times 10^{-10}$ determined from primordial abundances (cf. [37] p 157), we can calculate the *present* baryon density to be

$$\rho_B = n_B m_N = 3.5 \times 10^{-31} \text{ g/cm}^3 . \tag{5.56}$$

As an added note we can emphasize that the number of photons vastly outnumbers the number of baryons,

$$n_\gamma = 2 \times 10^9 n_B. \tag{5.57}$$

As a result of this, photons effectively *did not* encounter baryonic matter or electrons from a very early time in the history of the universe.

# 5.17   Expansion Since Equality of Radiation and Mass

From the analyses of primordial nucleosynthesis (deuterium to lithium), which is sensitive to the temperature and densities of baryons and radiation in the early universe, it has been found that $\rho_B \approx 3.5 \times 10^{-31} \text{g/cm}^3$. Baryon conservation has preserved this value in the intervening time.

The density of radiation can be found from the Stefan–Boltzmann law, $\rho_r = aT^4/c^2$. From the present temperature of the relic cosmic background radiation, 2.728 K$^o$ , we find $\rho_r \approx 4.7 \times 10^{-34}$ g/cm$^3$ (Section 5.16). From the above two densities the ratio of baryonic mass density (visible and dark) to radiation is

$$\rho_B(t_0)/\rho_r(t_0) \approx 750 \,. \tag{5.58}$$

The cosmic background radiation (CBR) is essentially the total mass density carried by radiation. The stars contribute only

$$\rho_*/\rho_B \approx 3 \times 10^{-5} \,. \tag{5.59}$$

By how much has the universe expanded since the radiation and baryon mass densities were equal? (Radiation photons do not have mass. But because of the Einstein equivalence of energy and mass, $E = mc^2$, a photon of frequency $\nu$ or equivalently wavelength $\lambda$ has a mass equivalent of $h\nu/c^2 = h/c\lambda$.) Matter density and radiation scale with the expansion differently, as already noted. So the following two relations can be written:

$$1/R(t_0) \sim \rho_r(t_0)/\rho_B(t_0) \,, \tag{5.60}$$

where $t_0$ denotes the present cosmic time, and

$$1/R(t_E) \sim \rho_r(t_E)/\rho_B(t_E) \equiv 1 \,, \tag{5.61}$$

where $t_E$ denotes the time at which the mass density of all baryonic matter and radiation were equal. Dividing the second by the first it follows that

$$R(t_0)/R(t_E) = \rho_B(t_0)/\rho_r(t_0) \approx 750 \,, \tag{5.62}$$

Thus the universe has expanded by 750, nearly a thousand fold, since matter first became dominant.

The wavelength of all radiation stretches in proportion to the universal expansion, and the temperature therefore falls with expansion,

$$\lambda \sim R, \quad \text{and} \quad T \sim 1/R \,.$$

So at that earlier time when radiation and matter were equal in mass density the temperature must have been larger than the present 2.7 degrees,

$$T_E = 2.7 \times 750 = 2000 \text{ K} \,. \tag{5.63}$$

## 5.18   Helium Abundance

Stars, even though they synthesize helium, cannot account for much of the helium in the universe because they burn it to produce heavier elements, or as in low mass stars like our Sun, the helium is locked within them for 12 billion years or more, aside from a little that is expelled into space by a wind from the flaming surface. Supposing that all the helium-4 observed in the universe were actually made in stars, then the equivalent mass density would have to be greater than 0.002 times the density of mass in the galaxies. But the actual ratio of all starlight to matter is not greater than 0.00003. (see ref. [37] p 159)

## 5.19   Helium Abundance is Primeval

It can be made quite clear that only a small fraction of the cosmic abundance of helium-4 can be synthesized in stars.

Each formation of a $^4He$ nucleus releases a binding energy of 27 MeV while the mass of a $^4He$ nucleus is 3728 MeV. Almost all of the binding energy makes the light by which stars shine. So the radiation produced per $^4He$ is

$$\rho_*/\rho_{\text{He}} = 27/3728.$$
(5.64)

From the observed abundance of $^4He$,

$$\rho_{\text{He}} \sim (1/4)\rho_m,$$
(5.65)

where $\rho_m$ is the estimated mass density of matter in the universe as determined by weighing and counting galaxies. In that case

$$\rho_*/\rho_m = (1/4)(27/3728) \approx 2 \times 10^{-3}$$
(5.66)

is the amount of radiation in starlight that we *should* see if all helium present in the universe today had been made in stars. But the ratio of the equivalent mass density of starlight to the density of matter is not more than $3 \times 10^{-5}$. So we must conclude that if the observed helium abundance were actually produced in stars, starlight would be $(2\times10^{-3})/(3\times10^{-4}) \approx 7$ brighter in equivalent mass than it is. Consequently, only a small fraction, if any, of the observed helium abundance could have been made in stars.

## 5.20   Redshift and Scale Factor Relationship

We seek the relation between redshift of radiation and the cosmic scale factor. Consider a far away galaxy from which we receive light. One crest arrives at $t_0$ and the next at $t_0 + \Delta t_0$ and they were emitted at $t_e$ and $t_e + \Delta t_e$ respectively $(t_0 > t_e)$. The light has traveled radially toward us on a nul geodesic of the (Robertson-Walker) metric

$$0 = \Delta t^2 - R^2(t)\Delta r^2/c^2(1 - kr^2). \tag{5.67}$$

From this we obtain

$$\int_{t_e}^{t_0} dt/R(t) = 1/c \int_0^{r_e} dr/\sqrt{1 - kr^2}, \tag{5.68}$$

and

$$\int_{t_e+\Delta t_e}^{t_0+\Delta t_0} dt/R(t) = 1/c \int_0^{r_e} dr/\sqrt{1 - kr^2}. \tag{5.69}$$

Because the right sides of both equations are equal, so too are the left sides. Therefore,

$$\int_{t_e}^{t_0} dt/R(t) = \int_{t_e+\Delta t_e}^{t_0+\Delta t_0} dt/R(t). \tag{5.70}$$

But for any frequency of electromagnetic radiation, the intervals between crests, $\Delta t_e$ and $\Delta t_0$, are fractions of a second, during which the relative distance the galaxy has moved is negligible. Therefore,

$$\Delta t_0/\Delta t_e = R(t_0)/R(t_e) > 1. \tag{5.71}$$

So the redshift (the fractional change in wavelength) is

$$z \equiv \lambda_0/\lambda_e - 1 = R(t_0)/R(t_e) - 1. \tag{5.72}$$

Thus we have the redshift $z$ in terms of the relative change in scale factor.

The redshift is often used by cosmologist in place of time because its relationship to the cosmic scale factors at two different epochs has the above direct relationship, whereas the actual time difference between the two scale factors depends on the cosmological constants which were at one time highly uncertain. Besides, the cosmological redshift is directly measurable.

## 5.21   Collapse Time of a Dust Cloud

We roughly estimate the collapse time of a spherical cloud of dust (no pressure) as follows. Equate gravitational potential energy of the outer shell of mass $\Delta m$ to an average kinetic energy,

$$GM\Delta m/r = (1/2)\Delta m\, v^2 \tag{5.73}$$

to get

$$\tau = r/v = \sqrt{3/(8\pi G\rho)} \sim 1/\sqrt{G\rho}. \tag{5.74}$$

## 5.22   Jeans Mass

In an otherwise uniform universe of hydrogen and helium, imagine a slight lumpiness here and there. Focus on one of dimension $R$. As it begins to collapse under its own gravity and its pressure rises, it may bounce, re-expand, collapse, bounce and so on. This oscillation has a period $R/v$, where $v$ is the velocity of sound in the clump. If the period is greater than the characteristic time for the collapse (estimated in the previous Section) the clump will collapse; otherwise it will oscillate.

Therefore, gravitational collapse can occur only if the following condition is satisfied,

$$R/v > 1/\sqrt{G\rho}. \tag{5.75}$$

From the value of $R$ given by this condition, we find the volume of the cloud, and from its density we obtain a mass that is referred to as the Jeans mass,

$$M_J = 4\pi\rho v^3/3(G\rho)^{3/2}. \tag{5.76}$$

## 5.23   Jeans Mass in the Radiation Era

The equivalent matter density of radiation on which gravity acts is

$$\rho_r = aT^4/c^2. \tag{5.77}$$

The pressure exerted by the radiation resists gravitational collapse, so we need to find under what circumstances of density and temperature gravity wins.

We can find the pressure of the radiation by noting first that $1/3$ of it is moving in any particular direction with speed of light $c$. The momentum of this $1/3$ of the radiation in a unit volume is $\rho/3 \times c$. Pressure is the momentum striking unit area per unit of time so that the radiation pressure is $(\rho/3 \times c) \times c$, or

$$p = \rho c^2/3. \tag{5.78}$$

Now we can find the velocity of sound in the cloud from

$$v = \sqrt{dp/d\rho}. \tag{5.79}$$

These results can now be used in the expression above to find the Jeans mass in the radiation era;

$$M_J = \rho c^6/[(3T^6(3Ga)^{3/2}]. \tag{5.80}$$

This mass is made up mostly of radiation, which will eventually escape from the protogalaxy. So we want to know the mass of matter that it contains.

The condition that matter and radiation where equally dense in mass, $\rho_m(t_R) = \rho_r(t_E)$, defines an era $t \leq t_E$, the era of radiation dominance, during which the *matter* density can be written in terms of the *radiation* temperature using relations derived in section5.10;

$$\rho_m(t) = \rho_m(t_E) \left( \frac{R(t_E)}{R(t)} \right)^3 = \rho_m(t_E) \left( \frac{T(t)}{T(t_E)} \right)^3$$

$$= \frac{a}{c^2} T(t_E) T^3(t).$$

We use this expression for the density of *matter* during the radiation era in terms of the radiation temperature at the time when the density of radiation and matter became equal. We obtain

$$M_J(t < t_E) = K/T^3 \sim t^{3/2}, \tag{5.81}$$

where

$$K = c^4 T(t_E)/3a^{1/2}(3G)^{3/2}. \tag{5.82}$$

## 5.24   Jeans Mass in the Matter Era

The gas that fills the universe in this era is approximately a monatomic gas of hydrogen having a specific heat ratio $\gamma = 5/3$. The pressure and sound velocity are given by

$$p = \rho_m k T_m / m_H, \tag{5.83}$$

and

$$v = \sqrt{dp/d\rho_m} = \sqrt{kT_m/m_H} \tag{5.84}$$

where $m_H$ is the mass of a hydrogen atom. Using these in the general expression for the Jeans mass we find,

$$M_J = (4\pi/3)(kT_m/Gm_H)^{3/2} \rho_m^{-1/2}. \tag{5.85}$$

We want to express the matter temperature and density, $T_m$ and $\rho_m$, in terms of the radiation temperature, $T$, alone. At times before and up to $t_E$, the matter and radiation temperatures are the same; this allows us to evaluate the constant in the condition for adiabatic expansion, $T_m V^{\gamma-1} =$ constant. We find

$$T_m(t) = T(t_E)/[\rho_m(t)/\rho_r(t_E)]^{2/3}. \tag{5.86}$$

Because, at all times, both in the radiation and matter eras, $T \sim 1/R$ and $\rho_m \sim 1/R^3$, we have (recall that $\rho_m(t_E) = \rho_r(t_E)$ by definition of $t_E$)

$$\rho_m(t) = \rho_r(t_E)[T(t)/T(t_E)]^{3/2}. \tag{5.87}$$

The last two results allow us to rewrite the Jeans mass in the matter dominated era as

$$M_J(t > t_E) = CT^{3/2} \sim 1/t \,, \tag{5.88}$$

with

$$C = \frac{4\pi}{3} \left( \frac{k}{Gm_H} \right)^{3/2} \rho_m^{-1/2}(t_E) \,. \tag{5.89}$$

## 5.25   Early Matter Dominated Universe

The formation of proto-galaxies and galaxies straddles the epoch when the universal expansion changed from being dominated by radiation to being dominated by matter. In section 5.10 we derived the way in which the universe expanded in the initial fireball. We turn now to the early part of the matter-dominated era. The cosmic expansion is different in the two eras for several reasons. First, the curvature constant $k$ and cosmological constant $\Lambda$ are not necessarily negligible as they were in the early fireball, although at *early* times of the matter dominated era before matter became very diffuse, they may be ignored. Second, the mass density of radiation and matter scale differently as the universe expands. Because of the Doppler shift of radiation, the radiation matter density scales as $1/R^4$ but matter density scales as $1/R^3$. In particular, for matter

$$\rho_m(t)/\rho_m(t_0) = R^3(t_0)/R^3(t) \,, \tag{5.90}$$

where $t_0$ is some convenient reference time, such as now, and $\rho_m$ refers to matter, not to radiation, which is now decoupled, and dilutes as $\rho_r \sim 1/t^2$ (section 5.10). However, the law that relates time and scale factor is different.

The Friedmann–Lemaître equation on page 190 becomes

$$\dot{R}^2 = (8\pi G\rho(t_0)R^3(t_0))/(3R(t)) \,. \tag{5.91}$$

Integrating from $t = 0$ to $t$ we find

$$R(t) = (6\pi G\rho(t_0))^{1/3} R(t_0)\, t^{2/3} \,. \tag{5.92}$$

From this we have also

$$\rho_m \sim 1/R^3 \sim 1/t^2 \,, \tag{5.93}$$

so that the *dominant* component of matter density scales with *time* in the same way in both the radiation and matter eras. For the temperature, it is different. The radiation temperature always scales as $1/R$ so that

$$T_{rad} \sim 1/t^{1/2} \,, \quad \text{before decoupling} \,, \tag{5.94}$$

whereas

$$T_{rad} \sim 1/t^{2/3}, \qquad \text{after decoupling}. \tag{5.95}$$

In contrast, the temperature of matter becomes ill defined for the universe later in the matter era, because matter condenses into clouds, galaxies, stars, and planets, each with its own temperature.

## 5.26  Curvature

Recall that in cosmology we deal always with events having a common cosmic time. They lie on a surface orthogonal to all the world lines at given time. By the cosmological principle the universe is homogeneous and isotropic. It follows that the *curvature* is everywhere the same else one observer would see things differently than another located elsewhere in the universe.

By definition, the Hubble parameter is

$$H \equiv \dot{R}/R. \tag{5.96}$$

The Hubble equation—the first of the pair of F-L equations—can be rewritten

$$kc^2/R^2 = H^2[\Omega_\Lambda + \Omega_M - 1] \tag{5.97}$$

where

$$\Omega_\Lambda \equiv \Lambda/3H^2, \qquad \Omega_M \equiv 8\pi G\rho/3H^2 \tag{5.98}$$

are dimensionless measures of the uniform mass (or energy) density represented by the cosmological constant $\rho_\Lambda = \Lambda/8\pi G$, and of mass density of all other kinds $\rho$.

If cosmological measurements should find that $\Omega \equiv \Omega_\Lambda + \Omega_M = 1$, then $k$ would have to be zero, and we would know that the universe is spatially flat. The meaning of $\Omega = 1$ can be deciphered by noting that

$$\Omega_\Lambda + \Omega_M = (8\pi G/3H^2)(\rho_\Lambda + \rho_M). \tag{5.99}$$

By defining a critical density

$$\rho_C = 3H_0^2/8\pi G = 7.7 \times 10^{-30} \text{ g/cm}^3 \tag{5.100}$$

(for $1/H_0 = 15 \times 10^9$ y) we see that flatness corresponds to there being a critical density $\rho = \rho_C$ in the universe.

$$\rho_\Lambda + \rho_M \equiv \rho_C \tag{5.101}$$

If this density is exceeded, the universe is closed. In fact we found on page 199 that $\rho_M \approx 3.5 \times 10^{-31}$ g/cm$^3$ Recent cosmological evidence points to $\rho_M \approx 2 \times 10^{-30}$ g/cm$^3$. This suggests that there is present in the universe additional *non-baryonic* matter of a so-far undiscovered nature. In either case, the universe is open with an accelerating expansion.

## 5.27   Acceleration

The second of the F-L equations dictates the value cosmic acceleration $\ddot{R}$. Until very recently (1999) it was believed that the gravitational attraction of the contents of the universe would decelerate the expansion. The acceleration equation can be rewritten using the earlier definitions as

$$q \equiv \ddot{R}R/\dot{R}^2 = \ddot{R}/RH^2 = \Omega_\Lambda - \Omega_M/2 - 4\pi Gp/H^2c^2. \qquad (5.102)$$

The last term—pressure—receives contributions only from relativistic sources like photons and neutrinos, which in the present and all recent epochs have small densities and even smaller pressure ($p = \rho c^2/3$). Matter, like baryons, and cold dark matter, even though they may dominate, are non-relativistic and have vanishing pressure. So for the dimensionless acceleration parameter we have

$$q \geq \Omega_\Lambda - \Omega_M/2. \qquad (5.103)$$

Notice that the curvature constant and the acceleration parameter define intersecting lines in the $\Omega_\Lambda$ vs $\Omega_M$ plane. So a measurement of curvature and acceleration will determine the values of the cosmological constant *and* the non-relativistic matter content of the universe. (Neutrinos are relativistic but not baryons.)

# References

[1] W. Baade and F. Zwicky, Phys. Rev. **45** (1934) 138.

[2] A. Hewish, S. J. Bell, J. D. H. Pikington, P. F. Scott and R. A. Collins, Nature **217** (1968) 709.

[3] J. R. Oppenheimer and G. M. Volkoff, Phys. Rev. **55** (1939) 374.

[4] N. K. Glendenning, *Compact Stars* (Springer–Verlag New York, 1'st ed. 1996, 2'nd ed. 2000).

[5] N. K. Glendenning, S. Pei and F. Weber, Phys. Rev. Lett. **79** (1997) 1603.

[6] N. K. Glendenning, Physics Reports **342** (2001) 393.

[7] N. K. Glendenning and S. Pei, Phys. Rev. C **52** (1995) 2250.

[8] E. Witten, Phys. Rev. D **30** (1984) 272.

[9] H. Minkowski, Ann. Phys. (Leipzig) **47** (1915) 927.

[10] J. D. Bjorken and S. D. Drell, *Relativistic Quantum Fields* (McGraw–Hill, New York, 1965).

[11] A. A. Michelson and E. W. Morley, Am. J. Sci. **34** (1887) 333.

[12] W. J. Swiatecki, Phys. Scripta **28** (1983) 349.

[13] W. J. Swiatecki, International J. of Modern Phys. E **15** (2006) 275.

[14] W. Voigt, Goett. Nachr. (1887) 41.

[15] A. Einstein, Ann. Phys., **49** (1916) 769.

[16] A. Einstein, *The Meaning of Relativity*, 5th ed. (Methuen and Co., London, 1951). A lecture series at Princeton University, 1921, with several revisions in later editions.

[17] S. Chandrasekhar, Astrophys. J. **74** (1931) 81.

[18] S. Chandrasekhar, Mon. Not. R. Astron. Soc. **95** (1935) 207.

[19] J. H. Taylor, A. Wolszczan, T. Damour and J. M. Weisberg, Nature **355** (1992) 132.

[20] C. M. Will, *Was Einstein Right?* (Oxford University Press, Oxford 1995).

[21] R. v. Eotvos, Math. Nat. Ber. Ungarn **8** (1890) 65.

[22] K. Nordvedt, Phys. Rev. **169** (1968) 1017; op. cit. **180** (1969) 1293; Phys. Rev. D **3** (1971) 1683.

[23] A. Einstein, unpublished manuscript in the Pierpont Morgan Library, New York (1920).

[24] R. V. Pound and G. A. Rebka, Phys. Rev. Lett. **4** (1960) 337.

[25] R. V. Pound and J. L. Snyder, Phys. Rev. Lett. **13** (1964) 539.

[26] J.C. Haefele and R.E. Keating, Science **177** (1972) 166.

[27] A. Schild, Texas Quarterly **3** (1960) 42.

[28] A. Schild, in *Evidence for Gravitational Theories*, ed. by C. Möller, (Academic Press, New York, 1962).

[29] A. Schild, in *Relativity Theory and Astrophysics* ed. by J. Ehlers (American Mathematical Society, Providence, R. I., 1967).

[30] J. Ishiwara, *Einstein Koēn-Roku* (Tokyo-Tosho, Tokyo, 1916). Record of Einstein's Kyoto lecture. The incident referred to is inferred by A. Pais to have occurred in November 1907.

[31] A. Pais, *Subtle is the Lord* (Oxford University Press, 1982).

[32] S. Weinberg, *Gravitation and Cosmology* (John Wiley & Sons, New York, 1972).

[33] M. Berry, *Principles of Cosmology and Gravitation* (Adam Hilgar, Bristol, 1989; 1st ed. 1976).

[34] A. Einstein and W. de Sitter, Proc. Nat. Acad. Sci. **18** (1932) 213.

[35] S. Perlmutter at al., Nature **391** (1998) 51; Physics Today (April 2003).

[36] A. Reiss et al., Astron. J. **116** (1998) 1009.

[37] N. K. Glendenning, *After the Beginning* (World Scientific Press (Singapore) & Imperial College Press (London), 2004).

[38] J. R. Oppenheimer and H. Snyder, Phys. Rev. **56** (1939) 455.

[39] M. A. Abramowicz and A. R. Prasanna, Mon. Not. R. Astron. Soc. **245** (1990) 720.

[40] M. D. Kruskal, Phys. Rev. **119** (1960) 1743.

[41] G. Szerkeres, Publ. Mat. Debrecen. **7** (1960) 285.

[42] J. B. Hartle in *Relativity, Astrophysics and Cosmology*, ed. by W. Israel (D. Riedel, Dordrecht, Holland, 1973).

[43] B. K. Harrison, K. S. Thorne, M. Wakano and J. A. Wheeler, *Gravitation Theory and Gravitational Collapse*, (University of Chicago Press, 1965).

[44] T. D. Lee, Phys. Rev. D **35** (1987) 3637; R. Friedberg, T. D. Lee and Y. Pang, Phys. Rev. D **35** (1987) 3640; ibid. 3658; T. D. Lee and Y. Pang, Phys. Rev. D **35** (1987) 3678.

[45] N. K. Glendenning, T. Kodama and F. R. Klinkhamer, Phys. Rev. D **38** (1988) 3226.

[46] A. S. Eddington, Observatory **30** (1920) 353.

[47] E. M. Burbidge, G. R. Burbidge, W. A. Fowler and F. Hoyle, Rev. Mod. Phys. **29** (1957) 547.

[48] S. A. Colgate and R. W. White, Astrophys. J. **143** (1966) 626.

[49] S. Bludman, D. H. Feng, T. Gaisser and S. Pittel, eds. *The Physics of Supernovae*, Phys. Rep. **256** (1995) 1.

[50] A. Burrows, E. Livne, L. Dessart, C.D. Ott, & J. Murphy, Astrophys. J. **640** (2006).

[51] A. Burrows and J. M. Lattimer, Astrophys. J. **307** (1986) 178.

[52] V. M. Kaspi, R. N. Manchester, S. Johnson, A. G. Lyne and N. D'Amico, Astron. J. **111** (1996) 2028.

[53] V. M. Kaspi, M. Bailes, R, N, Manchester, B. W. Stappers, J. S. Sandhu, J. Navarro and N. D'Amico, Astrophys. J. **485** (1997) 820.

[54] V. M. Kaspi, in *Neutron Stars and Pulsars*, ed. by N. Shibazaki, N. Kawai, S. Shibata, and T. Kifune (Universal Academy Press, Tokyo, 1998).

[55] V. M. Kaspi, Adv. Space Res. **21** (1998) 167.

[56] S. E. Thorsett and D. Chakrabarty, Astrophys. J. **512** (1999) 288.

[57] N. K. Glendenning, Phys. Lett. **114B** (1982) 392;  Astrophys. J. **293** (1985) 470;   Z. Phys. A **326** (1987) 57;   Z. Phys. A **327** (1987) 295.

[58] W. Keil and H. T. Janka, Astron. Astrophys. **296** (1995) 145.

[59] N. K. Glendenning, Astrophys. J. **448** (1995) 797.

[60] M. Prakash, J. R. Cooke and J. M. Lattimer, Phys. Rev. D **52** (1995) 661.

[61] I. Bombaci, Astron. Astrophys. **305** (1996) 871.

[62] M. Prakash, I. Bombaci, M. Prakash, P. J. Ellis, J. M. Lattimer and R. Knorren, Phys. Rep., **280** (1997) 1.

[63] J. R. Ipser, Astrophys. J. **158** (1969) 17.

[64] D. N. Schramm and R. V. Wagoner, Ann. Rev. Nucl. Part. Sci. **27** (1979) 37.

[65] S. M. Austin, Prog. Part. Nucl. Phys. **7** (1981) 1.

[66] D. Arnett, *Supernovae and Nucleosynthesis* (Princeton University Press, 1996).

[67] W. D. Myers and W. J. Swiatecki, Nucl. Phys. **A601** (1996) 141.

[68] N. K. Glendenning, Phys. Rev. D **46** (1992) 1274.

[69] R. A. Hulse and J. H. Taylor, Astrophys. J. (Letters) **191** (1974) L 59; op. cit., **195** (1975) L 51;  **201** (1975) L 55.

[70] J. H. Taylor and J. M. Weisberg, Astrophys. J. **345** (1989) 434.

[71] J. M. Weisberg and J. H. Taylor, Phys. Rev. Lett. **52** (1984) 1348.

[72] T. Kodama, Prog. Theor. Phys. **47** (1972) 444.

[73] Ya. B. Zel'dovich and I. D. Novikov, *Relativistic Astrophysics, Vol. 1, Stars and Relativity* (University of Chicago Press, 1971).

[74] A. Einstein, Ann. Phys., **35** (1911) 898.

[75] A. S. Eddington, Observatory **58** (1935) 37.

[76] S. Chandrasekhar, from the Eddington Centenary Lectures, 1982, reprinted in *Truth and Beauty* (University of Chicago Press, 1987) Ch. 6.

[77] K. R. Lang, *Astrophysical Data, Planets and Stars* (Springer-Verlag, New York, 1992).

[78] E. E. Salpeter and H. M. Van Horn, Astrophys. J. **155** (1969) 183.

[79] S. L. Shapiro and S. A. Teukolsky, *Black Holes, White Dwarfs, and Neutron Stars* (John Wiley & Sons, New York, 1983).

[80] L. I. Schiff, *Quantum Mechanics* (McGraw Hill Book Co., 1949).

[81] I. S. Gradshteyn and I. M. Ryzhik, *Tables of Integrals, Series and Products* (Academic Press, New York, 1965).

[82] P. M. Morse and H. Feshbach, *Methods of Theoretical Physics*, (McGraw-Hill, New York, 1953).

[83] L. D. Landau and E. M. Lifshitz, *Statistical Physics* (Addison–Wesley, Reading, Mass., 1969) Chapter VIII.

[84] N. K. Glendenning, Phys. Lett. **114B** (1982) 392.

[85] N. K. Glendenning, Astrophys. J. **293** (1985) 470.

[86] G. Baym, C. Pethick and P. Sutherland, Astrophys. J. **170** (1971) 299.

[87] P. J. Siemens, Nucl. Phys. **A141** (1970) 225.

[88] G. Baym, H. A. Bethe, and C. J. Pethick, Nucl. Phys. **A175** (1971) 225.

[89] R. H. Fowler, Mon. Not. R. Astron. Soc. **87** (1926) 114.

[90] L. D. Landau, Phys. Z. Sowjetunion **1** (1932) 285.

[91] S. Chandrasekhar, Phys. Rev. Lett. **12** (1964) 114; op. cit. p. 437.

[92] E. Schatzman, Astron. Zhur. **33** (1956) 800.

[93] B. K. Harrison and J. A. Wheeler, cited in B. K. Harrison et al., *Gravitation Theory and Gravitational Collapse* (University of Chicago Press, 1965).

[94] S. Chandrasekhar and R. F. Tooper, Astrophys. J. **139** (1964) 1396.

[95] T. Hamada and E. E. Salpeter, Astrophys. J. **134** (1961) 683.

[96] V. Weidemann, Annu. Rev. Astron. Astrophys. **28** (1990) 103.

[97] D. Koester, H. Shultz and V. Weidemann, Astron. Astrophys. **76** (1979) 262.

[98] L. L. Foldy, Phys. Rev. **83** (1951) 397.

[99] P. A. Seeger, Nucl. Phys. **25** (1961) 1.

[100] H. Nagara, Y. Nagata and T. Nakamura, Phys. Rev. A **36** (1987) 1859.

[101] N. K. Glendenning, Ch. Kettner, and F. Weber, Astrophys. J. **450** (1995) 253.

[102] H. A. Buchdahl, Phys. Rev. **116** (1959) 1027.

[103] R. C. Tolman, *Relativity, Thermodynamics and Cosmology* (Clarendon Press, Oxford, 1934).

[104] R. Geroch and L. Lindblom, Ann. Phys. **207** (1991) 394.

[105] H. Bondi, Proc. R. Soc. (London) **A282** (1964) 303.

[106] C.-S. Wang and W. Wang, Commun. Theor. Phys. (China) **15** (1991) 347.

[107] J. L. Friedman, J. R. Ipser and L. Parker, Phys. Rev. Lett. **62** (1989) 3015.

[108] P. Haensel and J. L. Zdunik, Nature **340** (1989) 617.

[109] Rev. John Michell, Phil. Trans. R. Soc. (London) **74** (1784) 35–57.

[110] J. E. McClintock, in *Physics of the Accretion onto Compact Objects*, ed. by K. O. Mason et al., (Springer-Verlag, Berlin, 1986).

[111] A. P. Cowley, Annu. Rev. Astron. Astrophys. **30** (1992) 287.

[112] J. Casares, P. A. Charles and T. Naylor, Nature **355** (1992) 614.

[113] F. Zwicky, Phys. Rev. **55** (1939) 726.

[114] C. E. Rhoades and R. Ruffini, Phys. Rev. Lett. **32** (1974) 324.

[115] M. Nauenberg and G. Chapline, Astrophys. J. **179** (1973) 277.

[116] L. Pietronero and R. Ruffini, in *Physics and Astrophysics of Neutron Stars and Black Holes*, ed. by R. Giacconi and R. Ruffini (North-Holland, Amsterdam, 1978).

[117] J. B. Hartle, Phys. Rep. **46** (1978) 201.

[118] J. L. Friedman and J. R. Ipser, Astrophys. J. **314** (1987) 594.

[119] J. L. Friedman, J. R. Ipser and L. Parker, Astrophys. J. **304** (1986) 115.

[120] F. Weber and N. K. Glendenning, Astrophys. J. **390** (1992) 541.

[121] J. M. Bardeen, K. S. Thorne, and D. W. Meltzer, Astrophys. J. **145** (1966) 505.

[122] M. Alford, K. Rajagopal, F. Wilczek, Nucl. Phys. B **537** (1999) 443.

[123] M. Alford, K. Rajagopal, S. Reddy, and F. Wilczek, Phys. Rev, D **64** (2001) 074017.

[124] K. Rajagopal and F. Wilczek, Phys. Rev. D **86** (2001) 1274.

[125] J. L. Zdunik, M. Bejger, P. Haensel, and E. Gourgoulhon, Astronomy and Astrophysics **450** (2006) 747.

[126] G. Baym and S. Chin, Phys. Lett. **62B** (1976) 241.

[127] G. Chapline and M. Nauenberg, Nature **264** (1976) 235.

[128] B. D. Keister and L. S. Kisslinger, Phys. Lett. **64B** (1976) 117.

[129] M. B. Kislinger and P. D. Morley, Astrophys. J. **219** (1978) 1017.

[130] B. Freedman and L. McLerran, Phys. Rev. D **17** (1978) 1109.

[131] W. B. Fechner and P. C. Joss, Nature **274** (1978) 347.

[132] F. Pacini, Nature **216** (1967) 567.

[133] P. Goldreich and W. H. Julian, Astrophys. J. **157** (1969) 869.

[134] J. P. Ostriker and J. E. Gunn, Astrophys. J. **157** (1969) 1395.

[135] M. A. Ruderman, in *High Energy Phenomena around Collapsed Stars*, F. Pacini, ed. (D. Reidel Publishing Company, Dordrect, 1987).

[136] B. R. Mottelson and J. G. Valatin, Phys. Rev. Lett. **5** (1960) 511.

[137] A. Johnson, H. Ryde and S. A. Hjorth, Nucl. Phys. **A179** (1972) 753.

[138] F. S. Stephens and R. S. Simon, Nucl. Phys. **A183** (1972) 257.

[139] R. N. Manchester and J. H. Taylor, *Pulsars* (W. H. Freeman, San Francisco, 1977).

[140] A. G. Lyne and F. Graham-Smith, *Pulsar Astronomy* (Cambridge University Press, Cambridge, 1990).

[141] J. B. Hartle, Astrophys. J. **150** (1967) 1005.

[142] N. K. Glendenning and F. Weber, Astrophys. J. **400** (1992) 647.

[143] N. K. Glendenning and F. Weber, Phys. Rev. D **50** (1994) 3836.

[144] N. K. Glendenning and S. A. Moszkowski, Phys. Rev. Lett. **67** (1991) 2414.

[145] N. K. Glendenning, *COMPACT STARS, Nuclear Physics, Particle Physics, and General Relativity* (Springer–Verlag New York, 1996).

[146] B. D. Serot and H. Uechi, Ann. Phys. (New York) **179** (1987) 272.

[147] J. I. Kapusta and K. A. Olive, Phys. Rev. Lett. **64** (1990) 13.

[148] J. Ellis, J. I. Kapusta and K. A. Olive, Nucl. Phys. **B348** (1991) 345.

[149] N. K. Glendenning, Nucl. Phys. B (Proc. Suppl.) **24B** (1991) 110.

[150] I. B. Khriplovich, Yad. Fiz. **10** (1969) 409;   D. J. Gross and F. Wilczek, Phys. Rev. Lett. **30** (1973) 1343;   H. D. Politzer, Phys. Rev. Lett. **30** (1973) 1346.

[151] N. K. Glendenning and C. Kettner, *Possible Third family of Compact Stars more Dense than Neutron Stars* Astron. Astrophys. **353** (2000) L9-12.

[152] M. C. Miller and F. K. Lamb, Astrophys. J. Lett. **499** (1998) L37-L40.

[153] M. C. Miller, F. K. Lamb, and D. Psaltis, *The Active X-ray Sky*, ed. by L. Scarsi, H. Bradt, P. Giommi, and F. Fiore, Nucl. Phys. B (Proc. Suppl.) **69** (1998) 123.

[154] S. W. Hawking, Nature **248** (1974) 30; Commun. Math. Physics **43** (1975) 199.

# Index

The author as taken by his daughter Elke May Glendenning in Marin County, California

## About the Author

Norman K. Glendenning took his bachelor's and master's degrees at McMaster University in Canada, and was awarded the Ph.D in theoretical physics by Indiana University in 1959. Glenn T. Seaborg invited him to continue his research at the Lawrence Berkeley Laboratory in 1958. There he has spent his entire professional career aside from frequent short-term visits of a month to a year as visiting professor in Paris and Frankfurt.

He has published 281 scientific papers in scholarly journals and in the proceedings of international conferences as invited lecturer—in North and South America, Europe, Asia, and India. As well, he has published a technical book in each of the two major fields in which he has done research over the years, **Direct Nuclear Reactions** (Academic Press), **Compact Stars** Nuclear Physics and Particle Physics and General Relativity, (Springer-Verlag).

More recently, he studies—as a student—the work of others, and writes what he has learned for the lay reader and scientists in other fields—**After the Beginning**: *A cosmic journey through time and space*. This book was selected by Scientific American Book Club as the alternate selection for July, 2005. The present book, **Our Place in the Universe** is his fourth book—again for the lay reader. Concurrently he is writing a technical book on General Relativity.

# ASTRONOMY AND ASTROPHYSICS LIBRARY

**Series Editors:** G. Börner · A. Burkert · W.B. Burton · M.A. Dopita
A. Eckart · T. Encrenaz · E.K. Grebel
B. Leibundgut . J. Lequeux . A. Maeder · V. Trimble

**The Stars** By E. L. Schatzman and F. Praderie

**Modern Astrometry** 2nd Edition
By J. Kovalevsky

**The Physics and Dynamics of Planetary Nebulae** By G. A. Gurzadyan

**Galaxies and Cosmology** By F. Combes, P. Boissé, A. Mazure and A. Blanchard

**Observational Astrophysics** 2nd Edition
By P. Léna, F. Lebrun and F. Mignard

**Physics of Planetary Rings** Celestial Mechanics of Continuous Media
By A. M. Fridman and N. N. Gorkavyi

**Tools of Radio Astronomy** 4th Edition, Corr. 2nd printing
By K. Rohlfs and T. L. Wilson

**Tools of Radio Astronomy** Problems and Solutions 1st Edition, Corr. 2nd printing
By T. L. Wilson and S. Hüttemeister

**Astrophysical Formulae** 3rd Edition
(2 volumes)
Volume I: Radiation, Gas Processes and High Energy Astrophysics
Volume II: Space, Time, Matter and Cosmology
By K. R. Lang

**Galaxy Formation** By M. S. Longair

**Astrophysical Concepts** 4th Edition
By M. Harwit

**Astrometry of Fundamental Catalogues**
The Evolution from Optical to Radio Reference Frames
By H. G. Walter and O. J. Sovers

**Compact Stars.** Nuclear Physics, Particle Physics and General Relativity
2nd Edition
By N. K. Glendenning

**The Sun from Space** By K. R. Lang

**Stellar Physics** (2 volumes)
Volume I: Fundamental Concepts and Stellar Equilibrium
By G. S. Bisnovatyi-Kogan

**Stellar Physics** (2 volumes)
Volume 2: Stellar Evolution and Stability
By G. S. Bisnovatyi-Kogan

**Theory of Orbits** (2 volumes)
Volume 1: Integrable Systems and Non-perturbative Methods
Volume 2: Perturbative and Geometrical Methods
By D. Boccaletti and G. Pucacco

**Black Hole Gravitohydromagnetics**
By B. Punsly

**Stellar Structure and Evolution**
By R. Kippenhahn and A. Weigert

**Gravitational Lenses** By P. Schneider, J. Ehlers and E. E. Falco

**Reflecting Telescope Optics** (2 volumes)
Volume I: Basic Design Theory and its Historical Development, 2nd Edition
Volume II: Manufacture, Testing, Alignment, Modern Techniques
By R. N. Wilson

**Interplanetary Dust**
By E. Grün, B. Å. S. Gustafson, S. Dermott and H. Fechtig (Eds.)

**The Universe in Gamma Rays**
By V. Schönfelder

**Astrophysics.** A New Approach 2nd Edition
By W. Kundt

**Cosmic Ray Astrophysics**
By R. Schlickeiser

**Astrophysics of the Diffuse Universe**
By M. A. Dopita and R. S. Sutherland

**The Sun** An Introduction. 2nd Edition
By M. Stix

**Order and Chaos in Dynamical Astronomy**
By G. J. Contopoulos

**Astronomical Image and Data Analysis**
2nd Edition By J.-L. Starck and F. Murtagh

**The Early Universe** Facts and Fiction
4th Edition By G. Börner

# ASTRONOMY AND
# ASTROPHYSICS LIBRARY